Nóra McNamara
Stephen Moore

GW01459551

Sustainable Livelihood Approach

Stephen Morse · Nora McNamara

Sustainable Livelihood Approach

A Critique of Theory and Practice

∅ Springer

Stephen Morse
Centre for Environmental Strategy
University of Surrey
Guilford
UK

Nora McNamara
Missionary Sisters of the Holy Rosary
Dublin 5
Ireland

ISBN 978-94-007-6267-1   ISBN 978-94-007-6268-8   (eBook)
DOI 10.1007/978-94-007-6268-8
Springer Dordrecht Heidelberg New York London

Library of Congress Control Number: 2013930172

© Springer Science+Business Media Dordrecht 2013
This work is subject to copyright. All rights are reserved by the Publisher, whether the whole or part of the material is concerned, specifically the rights of translation, reprinting, reuse of illustrations, recitation, broadcasting, reproduction on microfilms or in any other physical way, and transmission or information storage and retrieval, electronic adaptation, computer software, or by similar or dissimilar methodology now known or hereafter developed. Exempted from this legal reservation are brief excerpts in connection with reviews or scholarly analysis or material supplied specifically for the purpose of being entered and executed on a computer system, for exclusive use by the purchaser of the work. Duplication of this publication or parts thereof is permitted only under the provisions of the Copyright Law of the Publisher's location, in its current version, and permission for use must always be obtained from Springer. Permissions for use may be obtained through RightsLink at the Copyright Clearance Center. Violations are liable to prosecution under the respective Copyright Law.
The use of general descriptive names, registered names, trademarks, service marks, etc. in this publication does not imply, even in the absence of a specific statement, that such names are exempt from the relevant protective laws and regulations and therefore free for general use.
While the advice and information in this book are believed to be true and accurate at the date of publication, neither the authors nor the editors nor the publisher can accept any legal responsibility for any errors or omissions that may be made. The publisher makes no warranty, express or implied, with respect to the material contained herein.

Printed on acid-free paper

Springer is part of Springer Science+Business Media (www.springer.com)

Foreword

Ireland has a proud history in relation to development cooperation. This history commenced with the development work pioneered by Missionary orders which has been complemented and built upon by NGOs and by the Irish Government's own programme; Irish Aid. One of the Missionary orders with which I have had the pleasure of knowing well during my many years in Africa is the Missionary Sisters of the Holy Rosary (MSHR); the Missionary congregation that helped create the development organisation in Nigeria that is at the very heart of the story set out in this book. Health and education has been the core work of MSHR for decades in Africa. These were services that were much appreciated and for which there was an unending demand among the communities where it worked. I have to say I availed of these health services myself when I had a near death experience with malaria in Zambia in 1996! Thanks to Sr. Lucy and Dr. Vincent in Monze I am still of this world!

The Second Vatican Council (1962–1965) posed a clear challenge to those working in development such as MSHR which resulted in a greater and wider engagement with rural disadvantaged communities, primarily comprising peasant farmers. This was facilitated by a number of groups initially in Ireland, (particularly Gorta) that were keen to support novel ideas as a means of tackling the root causes of hunger, famine and disease—a new way of extending the hand of partnership and friendship.

But Ireland does not work alone in the development-sphere. Irish organisations partner with a host of other agencies, both national and international, as well as the private sector. We have a long history of working with our neighbours across the water in Britain and further afield. The Department for International Development (DFID) has provided much support for grass root development agencies in the Global South, including those from the Catholic Church. These days it is almost fashionable to be cynical concerning the value of development aid. Critics of aid single out and focus on examples of bilateral support provided to governments in the Global South who then mismanage or misappropriate the funds. It seems that all it takes are but a few examples of such bad practice to tarnish the whole of an aid programme.

Governments, including Ireland and Britain, and many others provide development assistance through a number of systems. This includes funding for grass roots organisations, most of them non-governmental in nature, with long histories of working at the level of communities. But working through governments, with the appropriate controls in place, ensures that work at grass roots levels is sustainable. I remember well from my days in Zambia, as I criss-crossed the country and visited many congregations living and working far from the capital Lusaka. The messages I brought back to the capitol were clear. We need government systems to work. Our schools need trained and motivated teachers, thoughtful curricula, and good management. These can only be provided by effective governments!

I know that MSHR has been involved in a number of research projects in Nigeria providing the centre-point—the plot as it were—for the story of this book. Gorta was also involved in supporting work related to those projects. It demonstrates how synergies in development can produce impacts greater than the sum of the parts, and here are the lessons learned vis-a-vis the Sustainable Livelihood Approach (SLA).

We in Irish Aid and indeed Misereor have been delighted to be effective partners contributing not only financially but more particularly with evaluations, discourse and advice that we hope gave rise to a form of true partnership that has become the heartbeat of the on-going but ever changing interventions. If any lesson was particularly learned it was one that reinforced the compelling need to ensure that beneficiaries took the lead (the "driving seat") in deciding how and what was done in all interventions. Development must be owned, locally and nationally by the individuals, communities and governments of the partner countries. Otherwise it is simply not sustainable. Over 35 years I understand that 350,000 people were reached. What a tremendous achievement.

It is my great pleasure to introduce this book and the work that the authors have carried out over many years. It is especially important that they share the knowledge gained in relation to what works in addressing what seems almost insurmountable poverty. There is no "magic bullet" to reducing poverty, addressing hunger and achieving "development". That is a conceit better left to the headline writers and the plethora of authors with ideologies or other solutions to peddle. Communities are complex and the causes of poverty are many. Hence interventions designed to help people have to work within this mosaic of components and interactions rather than pretend they do not exist. There have been many failures in development but let us not forget the numerous successes that have been and are still occurring. There is good news emanating from Africa and the statistics underpinning this news are available for anybody who cares to look.

The SLA was born out of a desire to ensure that progress in human well-being was lasting. At its heart lies a sensible preposition—that there is a need to understand the complexity of societies in order to make interventions on livelihoods optimally effective. The authors provide an honest and refreshing—warts and all—analysis of that process and set out some of the wider implications for development practitioners. There are always new challenges; that is why there is no 'end point' in development. Indeed the human race will continue to develop as long as it survives. Learning is intertwined with development—the latter cannot happen without the former. Excellent ideas have to be modified in the light of learning following experience. SLA is no different in that regard and it is my belief that the experiences set out in this book will help the development community become even more effective.

This publication will be of great value to all those interested in the 'doing' of SLA, be they development students, researchers, teachers or practitioners. That list includes organisations such as my own. While SLA can be seen as an approach to research, it was developed primarily as a means as a tool to support intervention. Moving from SLA (but still using it as a base) to the critical contemporary question of Sustainable Lifestyles gives much room for thought regarding individual and collective responsibility for sustainable living on this planet over the coming generations.

Brendan Rogers
Director General
Irish Aid
Department of Foreign Affairs and Trade
Dublin
Ireland

Preface

The authors of this book have much experience working in rural development in Africa. One of us—Nora McNamara—began an organisation called the Diocesan Development Services based in Idah at the conclusion of the Nigerian Civil War in 1970. Idah had been a frontline town in that war and DDS had a mandate for helping the rural poor, of which there were many, and it was inevitable that this would revolve around agriculture. Idah and its surrounds lacked any industry which could provide employment and what had existed was destroyed by the war. The vast majority of households in the area depended upon agriculture for their survival. The other author—Stephen Morse—first came out to Nigeria in 1980 and worked with Nora in DDS for much of that decade. In those days the fashion was very much 'Integrated Rural Development' (IRD); the bringing together of agriculture as a source of income alongside other services such as education and health care as well as infrastructure such as roads and water provision. Women featured heavily in IRD, not least because they were a significant source of income to support household livelihood. Major development donors such as the World Bank were heavily promoting and 'doing' IRD, and at the time was difficult to think of any other way or working.

IRD has long since evolved in many directions, not least because of the growth in urbanisation and a lessening of the focus on agriculture by many aid agencies during the 1980s and 1990s. Fashions changed and new themes such as good governance, accountability and stakeholder participation became dominant in aid agencies. These are certainly not incompatible with IRD, of course, but it is odd how fashions surge and wane and how lessons and experiences are forgotten. IRD had always been associated with 'projects'; a more micro-scale intervention planned for a discrete period of time and resource. Projects also began to wane as the focus shifted to programmes; longer term interventions. However, the central idea of IRD—bringing together many aspects of importance to the poor rather than focussing on but one—has shown resilience and does make a lot of sense. A sole focus on agricultural production without any consideration of how the produce would fare in markets where prices are unpredictable or the education of children or the supply of good quality water to help maintain the health of the household makes little sense. Even getting produce to markets requires roads. In the late 1990s a new form of integration came into being known as Sustainable

Livelihoods. It was an amalgam of many influences, and IRD was just one of a number of influences in this new wave, but sustainable livelihoods rapidly gained in popularity amongst development practitioners, researchers and policy makers. Sustainable livelihood has a strong focus on people and it is context neutral in the sense that it could apply to both rural and urban households, and the inclusion of the term 'sustainable' taps into a strong theme of making sure that what we do now does not damage future generations or restrict their livelihood choices. Sustainability was very much the theme of the 1990s and its popularity persists to this day, largely because like integration it makes a lot of sense.

This book is about sustainable livelihoods in practice, or more accurately the framework that was developed to operationalise it—the Sustainable Livelihood Approach (SLA). The authors have tapped into their long experience in working in development, but more importantly the long-lasting relationship they have had with the people in DDS. Nora has long since left the organisation she initiated but it does mean that she is well known to them and is trusted with their stories and secrets. Steve also worked for DDS for a time in the 1980s and 1990s, and the bonds are strong. This long-lasting relationship has allowed the authors to have access to the workings of DDS and in particular its attempt to make SLA work in the context of a series of interventions it was planning. The authors were somewhat on the periphery of all of that but were intrigued nonetheless by the ways in which SLA was put into action and, more importantly of all, the reasons why DDS did it and what they sought to gain. The process took two years and provided lots of opportunity for discussion and reflection between the authors and the DDS staff, the participants and the village communities as a whole. New ideas emerged and the authors became more and more determined to write them down and present them to a wider audience. The lessons of practice generated new knowledge and enhanced the wisdom that comes with years of experience and reflection, Indeed the experience and the outcomes were, at least to the authors, unique and not to be found in the SLA literature of which they were aware. This book is the authors' attempt to share those insights and they hope that the reader will find them useful. The SLA described in some detail may be rooted in one place and time but in the views of the authors they can transcend such borders and help to inform SLAs for other places and times. Unfortunately it can be all too easy to reject 'case studies' as being of little wider relevance but the authors feel that there are insights from the DDS experience that can 'travel'.

The authors would like to thank the staff of DDS for allowing such access and their permission to publish and share the findings and insights. They would especially like to thank Clement Agada, Moses Acholo, Gerlad Obaje, Remiguis Ikkah, Raphael Akubo and Stephen Okanebu. A special word of thanks to Chief Phillip Okowli and Chief John Egwemi who have travelled the DDS route with the authors since its inception; their insights and knowledge and love of their people was inspirational and their advice and good judgement contributed to good decision making. Reverend Sisters Monica Devine, Ruth Kidson, Madeleine Aiken, Denise McCarthy and Josie Burke must be thanked for their on-going interest and support in the work of DDS and to Nuala O'Donnell for reading the chapters and providing suggestions for improvement.

The SLA described in the book was funded via a number of sources including the Department for International Development (DFID) of the UK and Gorta (Ireland). The views expressed are solely those of the authors and do not necessarily reflect the views of DFID, Gorta or indeed DDS.

Acknowledgments

The authors would like to begin by acknowledging the support of the late Bishop of Idah Diocese, Dr. E. S. Obot, and for encouraging this type of intervention.

Within DDS, there are many who deserve mention including Sr. Felicitas Ogbodo Remigius Ikkah, Moses Ocholo, Gerard Obaje, Raphael Akubo, Clement Agada, Sunday Onakpa, Stephen Okanebu. The authors would like to thank DDS for allowing access to their work and their patience, interest and enthusiasm shown while this research was in progress.

A number of key informants helped with this project and we thank them all for their time and insights. The authors would especially like to thank Chief Phillip Okwoli, Chief John Egwemi and Mr. Stephen Okanebu.

Stephen Morse would like to thank the Universities of Reading and Surrey for providing him with the time and space to do the research that forms the basis of the book as well as the writing. He would especially like to thank Professors Rob Potter, Stephen Nortcliff and Mathew Leach.

Nora MacNamara wishes to acknowledge in a special way Nuala O'Donnell, Dorothy Shevlin and Madeleine Aiken for their editorial comments based on their extensive experience in Africa and for their encouragement to continue in this work.

Contents

Abbreviations

AADP	Anyigba Agricultural Development Project
ADP	Agricultural Development Project
BRIC	Brazil, Russia, India and China
CRP	Conservation Reserve Program
CSN	Catholic Secretariat of Nigeria
CU	Credit Union
CWO	Catholic Women's Organisation
CYON	Catholic Youth Organisation of Nigeria
DDS	Diocesan Development Services
DEFRA	Department for the Environment, Food and Rural Affairs
DFID	Department for International Development
EG&S	Ecosystem Goods and Services
EU	European Union
FC	Farmer Council
FEED	Farmers Economic Enterprise Development
FSR	Farming Systems Research
GDP	Gross Domestic Product
HDI	Human Development index
HDR	Human Development Report
HH	Household
HHH	Household Head
IFAD	International Fund for Agricultural Development
IMD	Index of Multiple Deprivation
IMF	International Monetary Fund
IRD	Integrated Rural Development
MOA	Ministry of agriculture
N	Sample size or Naira (Nigerian currency)
NDP	National Development Plan
NGO	Non-Governmental Organisation
NRC	National Republican Convention
PDP	People's Democratic Party
PES	Payment for Ecosystem Services

PRA	Participatory Rural Appraisal
RRA	Rapid Rural Appraisal
SAP	Structural Adjustment Programme
SD	Standard deviation
SDP	Social Democratic Party
SE	Standard Error
SLA	Sustainable Livelihood Approach
SLifA	Sustainable lifestyle analysis
SSP	Single Super Phosphate
SULA	Sustainable Urban Livelihoods Approach
TID	Townsend Index of Deprivation
UK	United Kingdom
UNCED	United Nations Conference on Environment and Development
UNDP	United Nations Development Programme
UPE	Universal Primary Education
USA	United States of America
WCED	World Commission for Environment and Development

Chapter 1
Sustainability and Sustainable Livelihoods

1.1 The Future of Sustainability

Sustainability is one of the defining words of our age. It implies a sense of longevity—something that will last well into the future—and as a consequence it implies a resilience to the turbulence of our politics, economic systems and environmental change that seems to be so embedded within our world. These are powerful ideas. Who wouldn't want to make sure that our generation doesn't despoil the planet for future generations? Who would want to cheat on their kids? Even the most diehard climate change sceptic would hardly wish to destroy the habitability of the planet. Their issue is much more with the evidence that underpins claims that climate change is 'real' than it is with the central ethic that making the world inhabitable for human beings is undesirable. Given that sustainability means continuation into the future then it is perhaps no surprise that it has been around as long as the human race has been able to think about the future. The Nobel Prize willing biologist Gerald Edelman regards consciousness as being of two types (Edelman 2003), and in simple terms these are:

1. Primary. An ability to recognise ones present and recent past and makes links between them; a remembered present.
2. Secondary. An ability to go beyond what is included under 'primary' by thinking about the future; the plans an individual may have. Thus "Higher-order consciousness allows its possessors to go beyond the limits of the remembered present of primary consciousness. An individual's past history, future plans, and consciousness of being conscious all become accessible. Given the constitutive role of linguistic tokens, the temporal dependence of consciousness on present inputs is no longer limiting" (Edelman 2003; 5521–5522).

Admittedly this is a biological vision of consciousness as it is clearly linked to evolutionary advantage. Hence the first of these clearly brings with it a competitive advantage as animals can draw upon a remembered immediate experience to help handle a current challenge. Secondary consciousness is a major step

S. Morse and N. McNamara, *Sustainable Livelihood Approach*,
DOI: 10.1007/978-94-007-6268-8_1,
© Springer Science+Business Media Dordrecht 2013

further than this as animals can make sense of a complex challenge not previously experienced. Self-recognition is often said to be a critical aspect for secondary consciousness and few species outside of humans have been shown to posses it. But while there is a further evolutionary advantage in developing a secondary consciousness the future based upon a set of actions may not necessarily be totally predictable. Many actions do provide a predictable outcome; throw a stone in the air and it will come down. But other actions can create multiple and uncertain futures, even if some are more likely than others. Thus it is certainly true that there are some, including industrialists, scientists, politicians and business people who are beginning to question evidence that links some human activity to an inevitable and negative environmental impact. They argue that there is some ambiguity in predicted outcome, even if science suggests that there are some outcomes much more likely than others. Such interrelationship between what people do and the environment is a central aspect of sustainability; that what we do now should not cause harm to future generations. It is, for all intents and purpose, that sustainability is the peak realisation of secondary consciousness as it projects the outcomes of action far into the future. But decisions made now in a multitude of endeavours can have ramifications for economics, social change as well as the environment, and herein rests the 'non trivial' challenge for *Homo sapiens*. On the surface it might seem extraordinary that such heated debates can exist over the human-induced increase in atmospheric greenhouse gases and impacts on climate. In part this debate exists because of the uncertain (in the sense of being 100 % confident) prediction that comes out of the scientific evidence and in part because of the quite different vested interests from some of those involved. But if the doubting Thomas's get it wrong then in the worst case scenario the human race could quite literally become extinct—so much for the evolutionary benefits of secondary consciousness.

The world has become so accustomed to fashions; the buzz words that gain popularity but disappear once leaders change and the world moves on. Is sustainability in this category? That it is the buzz word of our age no one will deny but will it too have a specific life span—in other words is the concept of sustainability sustainable? The notion of a concept that is meant to encapsulate the notion of 'lasting' and 'enduring' being condemned to the dustbin of history must surely be one of the ultimate ironies. At one level it is understandable that sustainability has found itself under attack. Not only do we have experience of ideas (and terms) becoming fashionable and then unfashionable but in the case of sustainability it can be viewed as a threat to our consumer society; an attack on the business as usual culture. Sustainability can thus be seen as a brake or obstacle; something that makes you think too much about what is happening now or indeed provides an awkward set of checks and balances. To some this is anathema as it can damage their present quality of life. They may argue that the planet is a naturally resilient place making it possible for our present day unscrupulous unthinking activities to be easily mopped up by Mother Nature. Some also see the term as being far more about the environment and not enough about our lives. People can't resonate with it (so they say) because it doesn't matter to their day-to-day existence.

Most just want to survive (first) and prosper (second). Indeed the immediate horizon for many is to live to the next day and the future could almost be a different planet. Even for the more fortunate there are concerns such as keeping ones job, maintaining a life style, finding somewhere to live, educating children and dealing with such matters occupy most hours of each day. It is against this background that some have suggested replacing sustainability with a term that better resonates with the lives of many, such as maintaining/enhancing 'quality of life' or 'wellbeing'. Thus it is perhaps ironic that one of the problems of sustainability is that in the eyes of many it has become too associated with the environment—it is almost too green. The centrality of people in sustainability can easily be lost, yet sustainability is all about people. Indeed it is here where the central theme of this book—livelihoods—resides. Livelihood, of course, is a people-centred concept with the environment important for people's livelihoods and indeed survival—a critical aspect of the mix. But there are other factors too in this mix for a sustainable livelihood besides the environment as sustainability comprises a multiverse of concerns.

1.2 The Multiverse of Sustainability

Given that sustainability is about us, it inevitably embraces and takes into consideration all that we do, and by 'us' is meant everyone on the planet. While many often equate sustainability with a concern for the environment it is so much more. Trying to capture this within a diagram is not an easy task, even though the most commonly employed device is to see sustainability as the intersection of a series of three overlapping circles that symbolise the environment, the economic system and society. Figure 1.1a provides such an example of a sustainability '3 circles' diagram. This diagram does have problems as a conceptual device for representing sustainability. For example, note how in Fig. 1.1a the circles are presented as being of equal size and this is meant to show how the three concerns are of equal concern. Yet in practice while the need for the three circles may be recognised the relative size or importance of the circles within sustainability can change depending upon one's views. For example, some might regard the economic circle as the more important and so that circle would dominate the other two. In effect a reduction in the other two circles, but especially perhaps the environment, would be acceptable to allow the growth of the economic system. This is weak sustainability; an acknowledgement that the environment circle is required but an acceptance that it can be 'traded' against growth in the other two, but usually economic growth. Others would see the environment as being the most important as it is the basis for our survival and hence consider the size of that circle as being non-negotiable. This is what is known as strong sustainability. The polarity of weak and strong sustainability which involves the economic and environment circles appears to diminish the importance of the third circle (society), and indeed that circle is often the most under-represented in the sustainability discourse. This is perhaps

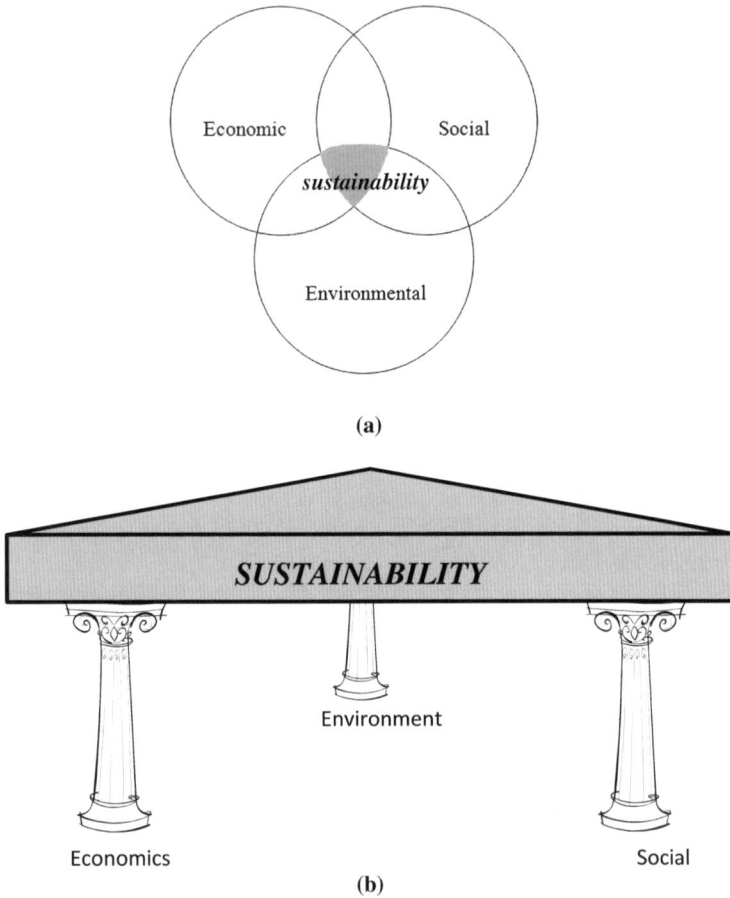

(a)

(b)

Fig. 1.1 The *3* circles or pillars of sustainability. **a** Sustainability as the overlap (*shaded*) between three circles. **b** Sustainability as a *3* cornered table supported by three pillars

understandable at one level given that conflicts typically arise between the need to generate economic capital and the often negative impacts that this has on our environment. Yet in the 19th century it was the clash between the generation of economic wealth and society which attracted the most attention. Karl Marx and Friedrich Engels, two of the most influential social thinkers of the time, sought to highlight the struggle between the upper and lower classes as the former sought to increase its economic capital via the exploitation of the latter. This 'economic capitalism' or 'capitalism' for short was regarded by Marx as unsustainable (although he did not use that term) and he predicted that the working class would eventually succeed in instilling socialism and ultimately the result would be a classless society where everyone was truly equal. One way of looking at this is a move from strong sustainability focussing on the 'social' to a strong sustainability focussed on the 'environment, with the drive towards amalgamating economic capital as the antithesis of each. This change in perspective helps to highlight a second issue with Fig. 1.1a; it evolves as our societies evolve. Our lives and indeed aspirations change as a result of many factors, such as globalisation, technology, the media and so on, and what comprise the circles as well as their importance evolve as a result.

An alternative conceptualisation of sustainability shown in Fig. 1.1b is with the three facets of sustainability presented as pillars of a three cornered table rather than overlapping circles. The assumption here, of course, is that if any of the pillars are removed then the table falls. This seems to have a better 'feel' of what sustainability is about relative to its being presented as a mere overlap of interests, but still does not get around the problem that the pillars may not necessarily need to be of equal size. Thus, some argue, the table may still be propped up with a strong economic pillar and weaker, but adequate, environmental and social pillars. As long as the table doesn't fall then all is well.

Clearly there is a limit to the imagery that can be used to present sustainability as both devices in Fig. 1.1 have their strengths and weaknesses. For all its faults, Fig. 1.1 does help to encapsulate the multiverse of concerns within sustainability and how they are inter-dependent, yet at first glance it also provides a daunting challenge for intervention. Indeed given the complexity in Fig. 1.1 and the fact that it changes over time it is perhaps not unsurprising that it comprises more a patchwork of interest groups promoting their own 'bit' of the diagrams than attempts to consider the whole. Even making predictions as to how this system will evolve is proving problematic. Of course this is not in itself unexpected. Marx for example suggested that the overthrow of capitalism and its replacement by socialism would be in the industrialised world but in fact it happened in the more agrarian-dominated economies of Russia and China. Marx based his prediction in part on what he regarded as an inevitable decline of the industrial working classes into poverty as a result of exploitation of their labour by the rich, but this failed to take into account changes in technology as well as the rise of liberal reform. Similarly the evidence-based predications of climate change as a result of human activities are denied by some who see this as a brake—a damper—on economic growth. But given the many challenges that humans face in basic requirements such as supplying adequate food and quality water for a growing population while at the same time ensuring that the planet does not become uninhabitable it appear there is little choice than to take all aspects of sustainability into account. But how can it be done?

Much has certainly been written about the meaning, origins and assessment of sustainability and it is not the intention of the authors to enter that fray here. The interested reader is referred to texts such as Bell and Morse (2008) and Morse (2010) that cover some of that territory, and there is an abundance of papers published in a variety of journals. But it can certainly be argued that the term is meaningless unless it can be put into practice, and it is here the challenges really begin. Action can be encouraged at all levels of society, from the individual to the nation state and indeed to global institutions such as the United Nations and

World Bank. At the smallest of these scales the phrase 'think global and act local' is often used with the understanding that if people can be encouraged to act in ways that encourage sustainability then this would 'scale up' and produce positive changes at the global level. With due deference to such considerations, this book is much more about the 'local' and its impact on sustainability than it is about higher scales; but the local is more than individual activity as it in turn can help to generate pressure on politicians, businesses and so on to also act for sustainability. After all, politicians depend upon votes and businesses upon customers. In due course, decisions made at higher scales of governance can influence what people do. Laws can be enacted that control actions, and policies can encourage or discourage certain activities. The larger and smaller scales are therefore inter-related.

1.3 Practicing Sustainability

One of the attempts to put sustainability into practice is to focus on livelihood. As noted above, this has the distinct advantage of connecting the concept to a notion with which people can resonate. Think of livelihood as a means of underpinning quality of life or wellbeing, and the connection is clear. Livelihood is not just about a means to survival but also about providing resources with which people can enhance and enjoy their lives. Rarely does one find the terms 'sustainability' and recreation and past times in the same paragraph let alone sentence, and this is regrettable. Unfortunately sustainability is often equated more with negative terms such as abstinence and austerity than with positives such as growth and enjoyment. So the term sustainable livelihood can be a means of making the connect between our day-to-day lives and the means by which we can sustain all this into the future without damaging any one else's prospects along the way. A refocusing of sustainability towards sustainable livelihood should provide the advantage of making it 'real' to people and exploring it in the light of people's lives does make a great deal of sense. It is our way and means of living that drives our consumption of the Earth's materials.

   Some appreciation of the increasing popularity of the term 'sustainable livelihood' can be gleaned from Fig. 1.2 which is a graph charting the use of the term within the topic and title of academic publications held on the ISI Web of Knowledge database. There are two parts to this graph, but the messages are alike. The first (Fig. 1.2a) is a simple count of the number of publications each year while the second (Fig. 1.2b) is an adjustment to allow for the number of pages in each article. The latter is designed to address any potential criticism that a simple count of articles can be inflated by short pieces such as book reviews or comments. Admittedly this analysis spans a somewhat limited group of specialised publications rather than the popular press, but even so it does provide an illustration with regard to a growing popularity amongst the academic and research community. Whichever way it is assessed, the use of the term within the topic of the publication has accelerated since the early 1990s, a period that included the

(a)

(b)

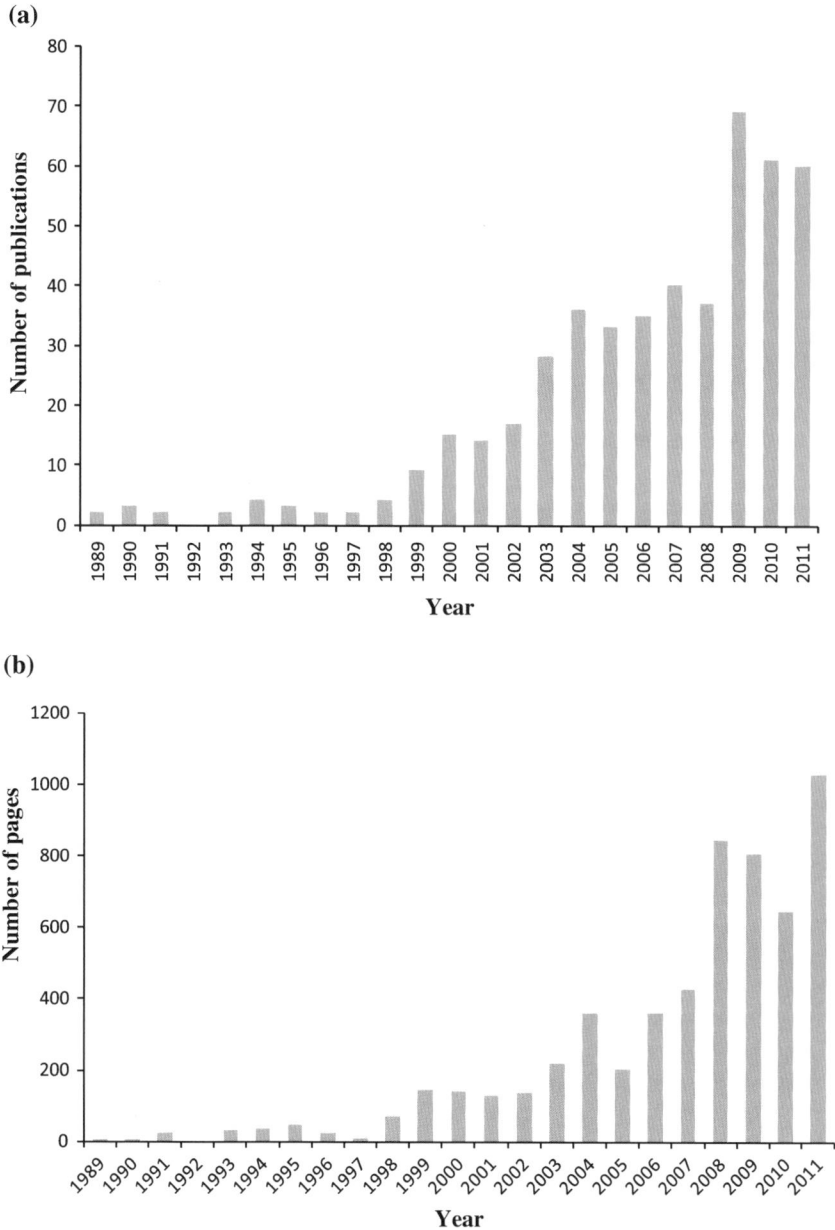

Fig. 1.2 Count of academic publications that include the term 'sustainable livelihood' in their title and/or abstract. **a** Number of publications. **b** Total number of pages in the publications in (a)

United Nations Conference on Environment and Development (UNCED) held in Rio de Janeiro between the 3rd and 14th of June 1992. This 'Earth Summit', as it is known, was a major event and attracted 108 heads of state. It spurred

the adoption of sustainability across the globe. In the recently (June 2012) held
'Rio+20' conference which took place 20 years after the Earth Summit the final
document agreed to by the attending Heads of State entitled *The Future We Want*
includes the statement:

> it is essential to generate decent jobs and incomes that decrease disparities in standards of
> living to better meet people's needs and promote sustainable livelihoods and practices and
> the sustainable use of natural resources and ecosystems.
> (The Future We Want; p. 5)

Indeed the term 'livelihood' appears no less than 12 times within the 49 page
document. It seems that the notion of a sustainable livelihood is not going to go
away, but the question is what is it? This question will form the theme of the next
chapter but at this point it is necessary to make a few observations that will appear
throughout the book.

Livelihood, of course, is a commonly employed word in English. The Oxford
Dictionary defines it as "a means of securing the necessities of life" and its origins are derived from an Old English word līflād 'way of life', from līf (meaning
'life') and lād (meaning 'course'). The 'hood' part of the word livelihood was a
later association. 'Hood' is used to denote a condition or quality, and also comes
from an Old English word, in this case hād (meaning 'person, condition, quality').
The word 'sustain' has its origins in the Old French word soustenir which in turn
is derived from the Latin word sustinere (sub = 'from below' and tenere = 'hold').
While both of the words have a long history and are now part of our everyday con-
versation, combining the two terms is a more recent event. Needless to say, even
the Oxford Dictionary definition of 'livelihood' is ambiguous. Just what are the
"necessities of life" that have to be secured?

Formal definitions of sustainable livelihood will be covered in Chap. 2, but
Fig. 1.3 sets out some of the icons—images that may well be evoked in the mind
of the reader when thinking about their "necessities of life". These icons have
not been linked to show consequence, but no doubt the reader can imagine how
some of them may be connected such that one needs another. At the centre there
is a classic icon for a household, defined in simple terms as a group of people
living together under the same roof. This may, or may not, be the same as family
but in this case it is; with a mother, father and two children. The main source of
income for the household may well be paid employment (yielding financial capital), perhaps on the part of one or both parents, and some of the earnings will
go to the government in the form of tax while part will pay for accommodation,
food, clothing etc. The rest may be saved or invested. Here the family livelihood
is founded upon paid employment, which in turn helps fund physical assets such
as a house, car and so on, as well as investment in children's education. The salary earned by the household will depend upon the skills and abilities of those in
employment (human capital), and included here may well be their professional
connections (social capital). But even this simple diagram raises a number of more
complex concerns. For example, how stable is paid employment and what may
influence that stability? Are there risks of redundancy? Will the government look
to increase tax take, and how and for what are those taxes used? The purposes

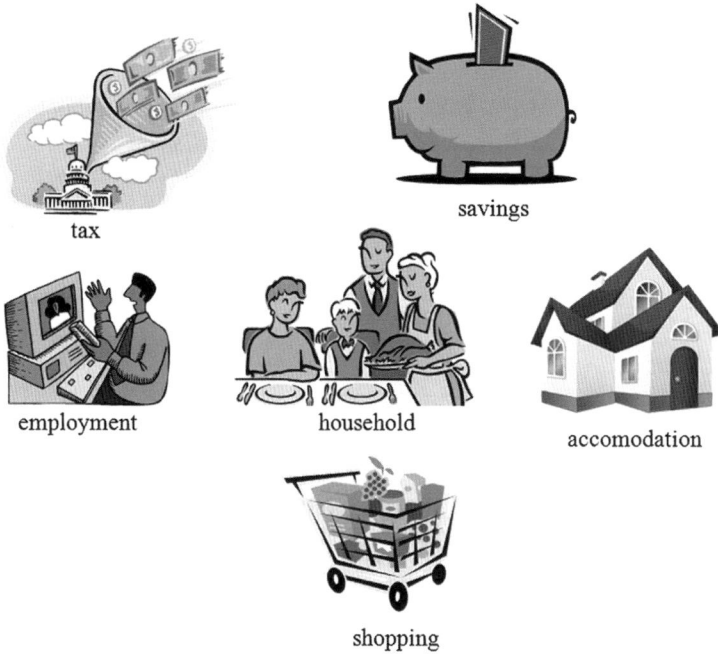

Fig. 1.3 Some icons of livelihood

for which the government uses that funding could indirectly help the household in ways that they can see; for example with investment in transport (roads, railways etc.), education and health care. But it may also be used to fund services which the household does not avail of in a direct sense (e.g. defence and foreign aid). Environment is absent from Fig. 1.3 as it is largely irrelevant in a direct sense for the family's livelihood. Perhaps it is only apparent to the household through weather which may restrict access to paid employment or purchasing of supplies. Extreme weather events may also cause damage to property. These have implications for livelihood and this in part may be influenced by a variety of institutional contexts such as the government and the employer as well as insurance companies (for damage to property). Even though there are hidden complexities within Fig. 1.3 it is nonetheless a representation of livelihood in one place; here it points to a typical developed world context where livelihoods are largely founded upon paid employment. It is thus a highly contextualised picture of livelihood—a vision that may dominate in richer parts of the world but not everywhere. In other places the components of livelihood may be quite different to what is seen in Fig. 1.3 and have nothing like the familiarity of icons for a 'western' reader. Livelihood may be far more diverse, with little paid employment for household members. Financial income may be based upon sales of farm produce, highly seasonal in nature and heavily dependent upon environmental and market conditions. Therefore the relative simplicity apparent in Fig. 1.3 can be replaced with something far more complex. Yet for someone born and raised within the context represented by the icons of Fig. 1.3 it can be challenging to recognise and appreciate such unfamiliar pictures of livelihood.

It follows from the preceding that any attempt to understand livelihood and per-haps intervene to make it more sustainable requires an approach that looks beyond the simple picture represented by Fig. 1.3. The essence may remain as households still have human, social, financial and physical capitals as these are broad terms, but what comprises them can vary as indeed can its institutional context and resil-ience. Logic would suggest that what is required is a framework that allows these to be taken into account before suggesting improvements; a framework that could be applied to people wherever and however, they may live. This should help ensure that any attempts made to intervene so as to improve the sustainability of liveli-hood would be based on evidence garnered from those meant to benefit rather than the whims and biases of those with the desire and power to impose. Put another way, such an approach facilitates a more evidence-based approach to intervention.

The Sustainable Livelihood Approach (SLA), the theme of this book, is, as one would expect from the title, all about people and their livelihoods. It is a people-centric way to the being and doing of sustainable development. This may appear to be counter to points made earlier about the need for the three circles/pillars of sustainability, but SLA it certainly does not try to discount and reduce the value of the natural environment within which this livelihood exists. Beyond this avowed focus on people what is so special about SLA that warrants a book such as this? Well it can perhaps be summed up using the words of the International Fund for Agricultural Development (IFAD) which regards SLA as both a "framework that helps in understanding the complexities of poverty" as well as "a set of princi-ples to guide action to address and overcome poverty". This sounds like an ambi-tious agenda—to not only understand poverty but also to help guide interventions to overcome it. Given such all-embracing, albeit somewhat vague statements, it is perhaps not surprising that SLA has obtained much popularity amongst develop-ment practitioners and researchers, and so should be the focus of many publica-tions designed to throw light on how SLA can be put into practice and lessons that have been learnt, than it has. This book is just part of that mosaic of experience. But while this sounds appealing what does this 'understanding' and 'action' entail exactly? IFAD sets out the principles of SLA as follows:

- *Be people-centred. SLA begins by analysing people's livelihoods and how they change over time. The people themselves actively participate throughout the project cycle.*
- *Be holistic. SLA acknowledges that people adopt many strategies to secure their livelihoods, and that many actors are involved; for example the private sector, ministries, community-based organizations and international organizations.*
- *Be dynamic. SLA seeks to understand the dynamic nature of livelihoods and what influences them.*
- *Build on strengths. SLA builds on people's perceived strengths and opportuni-ties rather than focusing on their problems and needs. It supports existing liveli-hood strategies.*

- *Promote micro–macro links. SLA examines the influence of policies and institutions on livelihood options and highlights the need for policies to be informed by insights from the local level and by the priorities of the poor.*
- *Encourage broad partnerships. SLA counts on broad partnerships drawing on both the public and private sectors.*
- *Aim for sustainability. Sustainability is important if poverty reduction is to be lasting.*

IFAD (www.ifad.org/sla/index.htm)

The reader can no doubt appreciate that this list covers much ground; spanning people, holism, dynamism and partnerships, let alone the need for sustainability. On the one hand it continues to be enticing as surely all these are indeed important for people, but it is also ambitious; it seems like a big ask to cover all in just one framework. Herein lies a conundrum with SLA. As both a framework and set of principles it says all the right things. It states that poverty is complex and for any intervention to succeed then it is necessary to understand what needs to be done and by whom. It says that people and institutions need to work together to maximise the chances of success. It also says that any attempt to improve matters for the poor must continue in terms of impact rather than be a once-off where the benefits disappear after a short time. Indeed it is impossible to criticise such core principles behind SLA, and this book will certainly not try to go down that road. After all, who can criticise the need to address poverty and improve the lot of many communities, probably still the majority of the world's population that needs help? SLA is founded on well-meaning principles and it is quite right to set out the breadth of issues needing to be addressed for interventions to work. The challenges with SLA are not in its mission and principles but with its practice. It is here difficulties arise, the reader may well expect this to be the case from a perusal of the list above and reflecting on how complex it would be to achieve any of these. Indeed, as this book will argue in later chapters, the SLA framework is, if anything, far too limiting and does not include enough.

The challenges involved in SLA will be set out in this book with reference to both the existing literature on the approach and also to a new and in-depth application of an SLA in a rural area of Nigeria. These are placed together in an approach that can perhaps be summed up as 'phronesis', a term used by Aristotle to represent practical wisdom or a third form of knowledge between theoretical and practical knowledge. The Nigerian case study helps to highlight points already made in the literature, but also highlights and adds other issues that the literature does not cover well. Perhaps the use of the 'case study' approach towards SLA in itself raises issues, including the wider applicability of findings that emerge. However, the geographical location of the SLA in Nigeria is typical of the 'place' where many SLA's have been done. Despite the fact that people everywhere have a livelihood, whether they be in Lagos or London, SLA has predominantly been applied within the context of a developing country by people based in the more developed world resulting in an in-built polarity from Global North (the so-called developed world) to Global South (the so-called developing world). In the early 21st century such labels are rapidly becoming blurred as countries such as Brazil, Russia, India and China (the BRICs) are fast developing economically even overtaking countries previously considered to be more 'developed'. But the polarity referred to for SLA still has some meaning, even if it too is blurred.

Much the same point regarding the 'place' of SLA could perhaps be said of the rural focus of the Nigerian case study. Many SLAs, probably the majority, have been applied in the rural context most likely because that is where many of the poor in the developing world endeavour to earn their livelihood. But there is nothing essentially 'rural' about SLA and it could be applied in any context where people have a livelihood. If the Human Race does manage to colonise other planets then in a thousand years from now an SLA could be applied to miners on the planet Mars or tourism centres on the Saturnian moon, Titan with its lakes of hydrocarbons. Wherever they may be, people always need a livelihood even if in the future the human race had outgrown the need for money and thus financial capital no longer existed. Indeed there are many examples of SLAs applied in peri-urban and urban contexts in the developing world. This is certainly not to say, that the results of applying an SLA in such quite different contexts would be the same—it is more than likely that they would not—but the people living in these places would still have a livelihood that can be analysed. Even so, the Nigerian case study is a rural one and the outstanding feature of it, which often emerges in SLA is the diversity of its components and the varied interactions that go to make up livelihoods. This is especially so in a country which has undergone rapid change in such a short space of time—the country was 'born' in 1960 and is younger than both authors of this book.

Finally, there is the reason why SLA is being implemented in the first place. The process is not just for learning purposes but should be geared towards action. In the previous paragraphs the action is vaguely articulated in terms of helping to address poverty, but just how is this to be done and by whom? Indeed what is meant by poverty in the first place and who defines it? There are scale issues here. Is the SLA meant to generate benefits within a relatively small scale—perhaps for a village or maybe a group of households within a village—or is it meant to feed into something bigger such as proposed changes in national policy? In the Nigerian case study covered in this book, the SLA was designed to feed into changes intended for a relatively small scale; one small organisation working in one small region of the country. That is fairly typical of many SLA studies as they are often applied at such small scales, often within the context of a single project. But it does raise questions as to applicability for changes planned at larger scales such as the nation state.

1.4 Structure of the Book

The book has four more chapters. Chapter 2 will provide some background of SLA; its origins, its promises and its practice. All are inter-related, as origins have helped shape the promise which in turn helped form the approach in practice. This causation is reversible. Practice in development over many years has helped shape what is needed to be known and thus helped to set out what is likely to be achievable as well as creating new ideas and theories as to what development means and how best to achieve it. In so doing, ideas constantly evolve and help drive new ways of practice. As will be shown in Chap. 2, SLA did not emerge from a vacuum; it evolved out of a series of approaches and ideas within development that evolved over decades.

Chapter 3 will set out some of the case-study context within which the SLA was applied. This includes the national (Nigerian) context but also the local one; the place where the SLA happened and indeed the institutions which facilitated it. Interestingly although as noted above the polarity of SLA has typically been developed (the 'doer' of SLA) to developing (the place of 'doing' the SLA) the case study introduces something of a novel polarity whereby the 'doer' of the SLA was a Nigerian institution. In this context 'doer' means those who thought of applying it, how it was to be applied as well as providing the resources for it to be done. Therefore it was not as such an SLA conceived and implemented by an outside (i.e. non-Nigerian) agency. Even with that caveat, the familiar North–South polarity was still apparent albeit in more subtle ways. Chapter 3 will cover this background and provide the reader with a 'feel' as to the important actors and agencies that were involved in the SLA and the context within which they worked.

Chapter 4 sets out some of the results of the SLA as applied in the Nigerian case study. The results have been summarised in Chap. 4 although they are still quite detailed. The reader's patience is requested at this point. The aim of the detail is to illustrate what can be found and the insights that emerge from an SLA and how these can be effectively utilised. While some of the material will be similar to those of other SLA studies there are also differences. This should be expected, as the findings of an SLA can be specific to the place where it is applied, but the case study also highlighted some more generic issues. The results will help provide the reader with a sense of how SLA works in practice and the practical difficulties that may be faced.

Chapter 5 will explore the lessons that emerge from the SLA literature and the case study, and use these to discuss what this means for the application of SLA and indeed the ideas upon which it is based. The North–South polarity of SLA will be returned to and this will help inform some of the theoretical failings behind SLA and how these can be addressed. The intention is certainly not to arrive at a 'better' SLA by tinkering around the edges but one which the authors think is more embracing of people's lives rather than just livelihoods. Indeed SLA is used to illustrate some of the key issues faced in sustainability.

Chapter 2
The Theory Behind the Sustainable Livelihood Approach

2.1 Introduction

This chapter will seek to set out the definition, origins and structure of SLA. In Chap. 1 it was pointed out that SLA is founded upon the notion that intervention must be based upon an appreciation of what underpins livelihoods. However there have been other factors at play that led to SLA as we know it today. First it is important to note that SLA was devised from what can be called an 'intentional' approach to development. Development has many meanings and Cowen and Shenton (1998) have made an interesting case for two basic forms:

1. Immanent development (or what people are doing anyway): this denotes a broad process of advancement in human societies driven by a host of factors including advances in science, medicine, the arts, communication, governance etc. It is facilitated by processes such as globalisation (an international integration) which helps share new ideas and technologies.
2. Intentional (or Interventionist) development: this is a focussed and directed process whereby government and non-government organisations implement development projects and programmes (typically a set of related projects) to help the poor. The projects are usually time and resource bound, but have an assumption that the gains achieved would continue after the project had ended.

Both of these forms can and do occur in parallel, with 'Immanent' development providing a broad background of change in societies while 'Intentional' development takes place as planned intervention. Thus, a country will be continuously undergoing 'Immanent' development as its public, private and 'Third' sectors gradually invest in infrastructure (roads, hospitals, water provision etc.), education and training, consumer products and services. The same country may also be host to a number of development projects, perhaps funded by foreign-based agencies. These project(s) may draw upon local expertise and resources, perhaps even secondments from public bodies, and may work in tandem with immanent development taking place in the country. Thus the national government may be

S. Morse and N. McNamara, *Sustainable Livelihood Approach*,
DOI: 10.1007/978-94-007-6268-8_2,
© Springer Science+Business Media Dordrecht 2013

investing in building and staffing of new hospitals, and a project may be funded by an international donor to help facilitate some aspect of this change. Similarly, the private sector may invest in new communication technologies such as a mobile phone network (immanent development) and in parallel a development agency may fund a project which explores how that new technology can be adapted to help with the delivery of a public service (intentional development). Projects within intentional development will typically have a 'blueprint' which sets out what has to be done, by whom and when, allied with some notion as to what the project is trying to achieve with the resources and time at the team's disposal. These objectives, methods and outcomes may be set out in formats such as a logical framework.

Immanent development has been around for as long as the human race but 'Intentional' development is a newer process. Indeed it can be argued that intentional development is largely a post—Second World War process that emerged from the 'Bretton Woods' institutions (named after the conference venue in New Hampshire where their creation was agreed). These institutions are best known as the International Bank for Reconstruction and Development (the World Bank) and International Monetary Fund (IMF). Both were born on 22 July 1944 and became operational in 1946. In the understandable optimism of those immediate post-war years a new president, Harry S Truman, came to power in the US following the death of Franklin D Roosevelt on 12 April 1945. President Truman won the next presidential election in 1948 in what is still regarded by many as the greatest election upset in American history. In the first national televised inauguration speech on January 20th 1949 he made the following statement:

> We are moving on with other nations to build an even stronger structure of international order and justice. We shall have as our partners countries which, no longer solely concerned with the problem of national survival, are now working to improve the standards of living of all their people. We are ready to undertake new projects to strengthen a free world. In the coming years, our program for peace and freedom will emphasize four major courses of action.

One of the "*major courses of action*" was set out as follows:

> we must embark on a bold new program for making the benefits of our scientific advances and industrial progress available for the improvement and growth of underdeveloped areas....... The United States is pre-eminent among nations in the development of industrial and scientific techniques. The material resources which we can afford to use for assistance of other peoples are limited. But our imponderable resources in technical knowledge are constantly growing and are inexhaustible. I believe that we should make available to peace-loving peoples the benefits of our store of technical knowledge in order to help them realize their aspirations for a better life. And, in cooperation with other nations, we should foster capital investment in areas needing development.

As well as optimism the speech also conveyed a sense of help and support for the poorer countries of the globe. One has to take care to put these intentions into the context of that era. This speech was delivered within a background of a growing 'Cold War' with the communist bloc, which in the coming years heralded much volatility and fear in the world. It is clear that the global engagement outlined in the speech was no doubt motivated in part by USA's self-interest to limit the international spread of communism, especially amongst the colonised countries of the Global South. The UK, Belgium and France, the predominant colonial countries of Europe, were on their knees economically and were being urged to hasten withdraw from their colonies in Africa and Asia. But many of their colonies were regarded by the USA as precisely those places where communism could flourish. The outbreak of the Korean War (1950) was only a few years away and there was growing unrest in Indochina (Vietnam). However, while the Truman speech is a convenient starting point for 'Intentional' development it is highly simplistic and perhaps unfair as it ignores what was happening before that year. For example, missionaries had long been engaged in 'intentional' development via the establishment of schools and hospitals.

Intentional development has had its fair share of critics, largely because it is based on a constructed sense of who is—and who isn't—developed and indeed what development actually means (Schuurman 2000). As highlighted with the Truman speech it tends to be the richer countries which set the agenda as to what needs to be done in the poorer countries. Escobar (1992, p. 413) for example regarded intentional development as nothing more than the "ideological expression of the expansion of post-World War II capitalism". Given the 'Cold War' context of the changes noted above then perhaps this should not be all that surprising, but it does help to highlight where the power rests with this process and it is an unequal distribution (Estreva 1992; Escobar 1992, 1995; Mathews 2004; Siemiatycki 2005; Simon 2006, 2007). Sidaway (2007) has even suggested that the practice of intentional development since the Second World War is almost a reconfiguration of colonialism as the rich, some of whom are old European colonial powers, dictate to the governments of their former colonies what they must do. Critics have also argued that Intentional development has by and large not been very successful (Rahnema and Bawtree 1997; Pieterse 1998; Hart 2001; Toner and Franks 2006), with Africa often cited as the classic example of failure (Mathews 2004). They point out that despite major investment by the developed world development projects have often failed to generate positive and sustainable outcomes for the people who were meant to benefit. These issues of visioning what is required for development and failure are not unrelated. If a vision of development from richer countries is being imposed in circumstances that are unsuitable then it is inevitably doomed to failure. As a result, there has been a backlash to such 'Intentional' development, often referred to as the 'post development' movement (Rahnema and Bawtree 1997) or sometimes more evocatively as anti-development (Simon 2006). It has to be noted that the post-development movement has had its own critics, largely because it can be quite woolly as to what can be done to help people living in poverty (Blaikie 2000). Some have even made the rather ironical point that post-development and capitalism have much in common as both appear to call for as little directed intervention as possible on the part of governments, albeit for entirely different reasons; i.e. they are both *laissez faire* in outlook.

SLA evolved within the context of the intentional development approach by which development practitioners were seeking to maximise the effectiveness of their interventions to help the disadvantaged. It is in effect a diagnostic tool which provides a framework for analysis leading to concrete suggestions for intervention (Allison and Horemans 2006; Tao and Wall 2009). It was typically applied in poorer countries as part of a planning phase for an intervention via policy, a development project or perhaps as the basis for more in-depth research. In that sense the SLA is an analysis of peoples' current livelihood and what is needed for an 'enhancement', and useful in avoiding the inappropriate interventions critiqued by the post-developmentalists. It should be noted that the latter might not necessarily be the need for people to replace their current livelihood or indeed have more means of livelihood. Instead it might involve making the current means of livelihood less susceptible to environmental, social or economic 'stresses'. The SLA could also result in recommendations that people themselves may be able to put into practice rather than be dependent upon the actions of outsiders. It is thus a 'no holds barred' approach to understanding and improving the sustainability of livelihood, although it clearly has to take into account what is feasible in different circumstances.

As set out here and in Chap. 1 it may be rather obvious to the reader that any attempt to improve livelihood should be founded upon an understanding of what is needed which must entail an appreciation of the diverse range of factors and processes that comprise livelihood? How can it be any other way yet still hope to succeed? It sounds so obvious. This intriguing question will be discussed later, but it is fair to say that 'integrated' approaches to 'Intentional' development did exist before SLA. Such integrated rural development projects (they were often based in rural areas) sought to bring together important components to development such as education, health, infrastructure and agriculture, which has some resonance with the 'integrating' basis of SLA. By way of contrast, it is also fair to say that historically many interventions geared towards addressing poverty tended to have a narrow perspective and were perhaps not 'joined-up' or not 'all embracing' (Krantz 2001). For example, poverty is not only about monetary income but has linkages to health and education as well as to perhaps less tangible entities such as a sense of 'powerlessness' (Krantz 2001). Thus poverty is multi-faceted, though the history of development suggests that a project ought to focus only on addressing one of the facets (income for example) and ignore all others. The project might have succeeded in boosting the income of some people but this might be at the expense of others, a boost that might be short lived. As noted above, the history of intentional development delivered via projects is a patchy one.

This chapter will explore some of the experiences to date with SLA. It will begin by setting out the nature of SLA, its definitions and origins, and move on to discuss the role of capitals and outline some of the critiques of SLA. The chapter will end with an import aspect of SLA that arguably has received less attention in the literature—how it can help translate new information to intervention. It should be noted that the sustainable livelihood, SLA and evidence-based intervention literature is a substantial one and the authors cannot claim to have mentioned every project, research and/or development in nature, where these aspects have played a role. Some of this has been reported in the academic literature but much also exists in so-called 'grey' form as reports residing within development funders, aid agencies, international agencies, consultancy companies etc. Some of this may be readily accessible while much may not be. Thus inevitably the chapter can only hope to skim the surface of this literature and the authors apologise in advance if the reader feels that a particular project or publication has been excluded.

2.2 The SLA Framework

The SLA framework is often formally set out diagrammatically as shown in Fig. 2.1 (Ahmed et al. 2011). An outline of SLA and suggestions for putting it into practice can be found in 'guidance notes' produced by DFID (available at www.nssd.net/references/SustLivel). Figure 2.1 is a far more sophisticated version of the collage in Fig. 1.3, and associated points made in the text of the previous chapter included in diagrammatic form. At its core is the assessment of the different capitals that are deemed to underpin livelihood at the level of the individual, household, village or group. These capitals are classified as human, social, physical, natural (a category not included in Fig. 1.3) and financial. They are then assessed in terms of their vulnerability to shocks and the institutional context within which they exist. Once this is understood then interventions can be put in place to enhance livelihoods and their sustainability, perhaps by increasing the capital available or by reducing vulnerability. Thus the process is about understanding the current situation and developing suggestions for improvement based upon that understanding. The SLA is meant to avoid a situation where intervention is unguided giving little positive impact or is at worst detrimental.

The reader will no doubt note that the SLA as set out in Fig. 2.1 is linear in style although in practice the interventions identified should give feedback to help

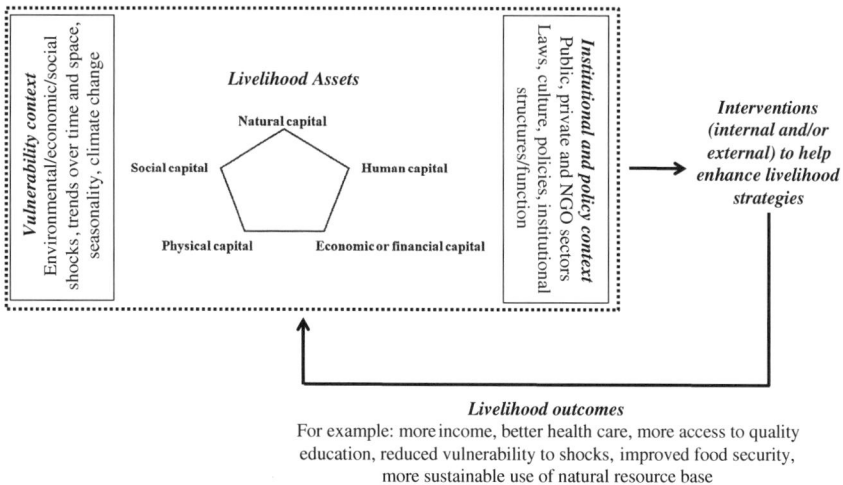

Fig. 2.1 The DFID sustainable rural livelihoods framework (after Carney 1998)

improve the capitals and contexts. It should also be noted that the degree to which this framework is meant to 'model' reality in any one case may be limited, but, of course, there has to be some semblance otherwise one could question the point of the exercise. The degree to which an SLA can generate an approximation to 'real' livelihood will be returned to later. Indeed it is also important to realise that the use of the SLA framework is not necessarily dependent upon facilitation by an external agency for those whose livelihood is being analysed. In theory it should be possible for anyone to apply this model. For example, an individual could apply it to themselves or to their household or a group could use it to analyse their own livelihoods. Also, SLA as set out in Fig. 2.1 does not specify particular methods and techniques that have to be applied to explore the capitals, institutions, vulnerability etc. In practice, the process of 'doing' an SLA could utilise a disparate range of methods including standard techniques based upon observation, focus groups and interviewing. The SLA is simply providing a framework as to what should be looked for and not necessarily how to do the looking.

However, while the logic behind the SLA has been set out here in a somewhat mechanical cause-effect terminology, it can be considered in many different ways. Krantz (2001, pp. 3 and 4) argues that there are two ways of using SLA. On the one hand there is the approach taken by DFID which sees SLA as a framework for analysis, while other agencies such as UNDP and CARE (an NGO) apply it to "facilitate the planning of concrete projects and programmes". The distinction made here would appear to be a rather fine one as the purpose of SLA is to help analyse a situation which would seem a logical fit with it's use in implement projects. Farrington (2001) presents a more nuanced view of the different dimensions of SLA:

1. As a set of principles guiding development interventions (whether community-led or otherwise). The fundamental assumption here is that an intervention has to be evidence-based rather than instigated in top-down fashion without adequate knowledge of the community. SLA can thus be seen as a loose checklist of points that need to be considered before an intervention is planned.
2. As a formal analytical framework to help understand what 'is' and what can be done. The framework helps aid an appreciation of the capitals which are available to households, their vulnerability and the involvement of institutions.
3. As an overall developmental objective. In this case development is seen as the improvement of livelihood sustainability, perhaps by making capital less vulnerable or by enhancing the contributions that some capitals can make or even by improving the institutional context.

The differences between these three dimensions of SLA may seem to be rather fine, especially with regard to 1 and 2. The Nigerian case study covered in later chapters will provide an illustration of the differences between these.

SLA has certainly helped establish the principle that successful development intervention, especially if led internally, must begin with a reflective process of deriving evidence sufficiently broad in vision and not limited to what may seem like a good 'technical' fix. This may be a surprising advance given that the logic upon which SLA is based seems clear—before development can take place there must be some idea what needs to be done, along with the why and what of how it must be done. It does imply a necessary degree of humility in that it suggests there is much to be learnt and understood before help is offered; this has to be built upon a partnership with those meant to benefit rather than seeing them just as passive recipients.

2.3 Definitions of SLA

SLA has been in vogue amongst development practitioners and researchers since the late 1990s and indeed was a central concept of the UK's Department for International Development's (DFID) strategy during the early years of the UK New Labour government. The call for an emphasis on sustainable livelihoods was set out in the 1997 White Paper on international development as follows:

> …refocus our international development efforts on the elimination of poverty and encouragement of economic growth which benefits the poor. We will do this through support for international sustainable development targets and policies that create sustainable livelihoods for poor people, promote human development and conserve the environment.DFID (1997: Summary, page 6).

What exactly are these 'sustainable livelihoods' that DFID intends to help create? Some illustration of this has already been provided in Chap. 1, but a definition has been provided by Chambers and Conway (1992) some five years before the White Paper:

> A livelihood comprises the capabilities, assets (stores, resources, claims and access) and activities required for a means of living; a livelihood is sustainable when it can cope with and recover from stress and shocks, maintain or enhance its capabilities and assets, and provide sustainable livelihood opportunities for the next generation; and which contributes net benefits to other livelihoods at the local and global levels and in the short and long-term.Chambers and Conway (1992, p. 7).

Assets are the same as the capitals mentioned earlier, but note how issues such as claims and access are included. An asset may not necessarily be owned by a household for it to be an important contributor to livelihood. As long as the household has access to it then it will help. Also, in this definition a number of strands coalesce. On one hand there is a requirement for a sustainable livelihood to be able to recover from "*stress and shocks*" but it must also be able to "*maintain and enhance*" capabilities and assets into the future. A central element in this 'resilience' to stress and shocks may well be the diversification of elements that comprise 'livelihood'. Hence a more diverse livelihood base could arguably be seen as more sustainable as shocks to one or more components can be compensated for by an enhancement of others. But this is conjecture and may not always be the case. A simplistic assumption that a diverse livelihood is more sustainable needs to be treated with caution.

Prior to publication of the White Paper, Carney (1998) provided a simpler vision of sustainable livelihood which has resonance with the definition of Chambers and Conway (1992):

> A livelihood comprises the capabilities, assets (including both material and social resources) and activities required for a means of living.

And, when merged with sustainability

> A livelihood is sustainable when it can cope with and recover from stresses and shocks and maintain or enhance its capabilities and assets both now and in the future, while not undermining the natural resource base.

The reader may be surprised that these definitions were not given earlier but in essence they are more concise and formal statements of points already made. What matters is that the definitions, for all their formality, are not surprising or unexpected. The simple example of a household livelihood provided in Chap. 1 has already set out the territory in terms that may well resonate with at least one group of readers; those born and raised within households where livelihood is largely dependent upon wage-earning. The definitions present the points discussed in a more generic way, but in more negative vein Carswell (1997, p. 10) has made the point that definitions of sustainable livelihoods are often "*unclear, inconsistent and relatively narrow*" and this could add to "*conceptual muddle*".

2.4 Origins of SLA

As already noted in Chap. 1, the notion of sustainable livelihood as we know it today can be said to have arisen out of the 1992 Earth Summit held in Rio (Perrings 1994) and its promotion of Agenda 21 (Agenda for the 21st Century). A stated aim in Agenda 21 is that everyone must have the "opportunity to earn a sustainable livelihood". Once the concept of a sustainable livelihood had been adopted then it seems like a small step to go from there to SLA. But SLA did not become main stream until the late 1990s, so why did the delay occur?

Like many initiatives in intentional development SLA did not come out of a vacuum nor indeed can it be said to have a definitive starting point. Rather it grew organically from a number of older trends and ideas; the term sustainable livelihood even predates the 1992 Earth Summit. For example there are influences arising from the application of 'systems' approaches to sectors such as agriculture. 'Agro- Ecosystem Analysis' has its origins in the 1960s and sought to bring together concepts in ecology along with social and economic aspects of agriculture (Conway 1985). These system-based (systemic) approaches were not just research frameworks but also had practical application. An example is the evolution of new approaches to knowledge generation with farmers. The historical approach had been to consider farmers as mere recipients of 'new' knowledge and technologies generated by research services and transferred via an extension service; hence the phrase 'transfer of technology'. Again the model was linear with information flowing one way. Newer systems changed this to a partnership approach towards knowledge generation, with farmers working together with researchers. Terms using the phrase 'farming systems' began to evolve in the 1980s to capture this new mentality; for example 'farming systems research' (FSR; Flora 1992). As one of the earliest papers on the application of systems thinking to natural resource management puts it:

> Systems or, to be more specific, systemic methodologies necessarily question the efficacy of linear models such as transfer of technology (TOT) and diffusion of innovations. Both neglect social and organisational processes in their assumptions about the nature of human communications and have been found wanting in many areas of rural development. Ison et al. (1997, p. 258)

Partnership approaches had to be built upon a genuine participation of farmers in the process; not a token representation where farmers were simply lectured to. Indeed FSR itself tapped into the parallel evolution of participatory methodologies since the 1960s (or indeed earlier) such as rapid rural appraisal (RRA) and participatory rural appraisal (PRA). Both RRA and PRA had a strong 'rural' focus (and exemplified in their respective names) and sought to include households in the knowledge generation process (Chambers 1991). RRA was more extractive in that it was intended as an umbrella term to cover a suite of methods by which researchers could learn about local livelihoods and so arrive at recommendations for intervention. PRA had the added thrust that potential interventions became part of the participatory-based discourse. This suite of methods used within PRA is much the same as those of PRA, and often used within SLA.

FSR, RRA and PRA are more focussed on work with households at village scale and as a result it is easy to see the resonance with SLA (Korf and Oughton 2006). All share a systemic mindset with a similar epistemology. However there are some resonances of the more macro-scale field of 'integrated rural development' (IRD), in vogue during the 1960s and especially the 1970s amongst major funders such as the World Bank (Yudelman 1976; D'Silva and Raza 1980; Krantz 2001). The literature on IRD is substantial and does not need to be reviewed in depth here. An early review of IRD which dates to the time when the concept was still popular is provided by Ruttan (1984). For recent discussions of successes/ failures the interested reader is referred to Gaiha et al. (2001) for IRD in India, Zoomers (2005) for IRD supported by the Netherlands 'Directorate-General for International Cooperation' (DGIS) and carried out between 1975 and 2005 in Asia, Africa and Latin America, and Fenichel and Smith (1992) for IRD in Zambia. The manifestation of IRD often took the form of large projects implemented over five years or so covering regions of a nation state with staff seconded from government agencies; a form of decentralisation. The 'integration' in the title usually meant a consideration of multiple sectors and how they interacted. Thus, it was argued that agricultural development also requires effective infrastructure such as roads to transport inputs and produce as well as adequate health care. The latter in turn depends upon good water supply in both quantity and quality. IRD projects were designed to address all these relationships and mark a break away from the older 'sector-specific' nature of development projects where agriculture (for example) may have been taken in isolation without any regard as to where inputs may come from or how farmers managed to get their excess production from enhanced yields to markets. Indeed if the word sector is replaced by asset or capital then

IRD would appear to have much in common with SLA. Figure 2.1 embodies this same sense of interaction.

Although SLA has resonance with older ideas one of its most prominent influences is the rise of what is referred to as 'human development' in the 1980s and promoted especially by the United Nations Development Programme (UNDP). Indeed SLA has been regarded by some as the 'operational vehicle' of human development (Singh and Gilman 1999). Human Development was influenced by the work of the Indian economist Amartya Sen and his writing on capability (Sen 1984, 1985) as well as other authors on vulnerability (Swift 1989; Chambers 1989; Davies 1996; Moser 1998) and access to resources (Berry 1989; Blaikie 1989). These are inter-related in the sense that having a more diverse capability can reduce vulnerability of livelihood to shocks in much the same way that biologists argue that greater biodiversity aids ecosystem resilience to shocks. 'Human development' took as its central tenet the importance of enhancing capability:

> Human development is a process of enlarging people's choices. In principle, these choices can be infinite and change over time. But at all levels of development, the three essential ones are for people to lead a long and healthy life, to acquire knowledge and to have access to resources needed for a decent standard of living. If these essential choices are not available, many other opportunities remain inaccessible.

UNDP Human Development Report (1990, p. 10)

Enlarging choices can be achieved by widening the capital base, for example by education and training. There are also nods in the direction of sustainable development albeit with an unambiguous focus on people:

> the development process should meet the needs of the present generation without compromising the options of future generations. However, the concept of sustainable development is much broader than the protection of natural resources and the physical environment. It includes the protection of human lives in the future. After all, it is people, not trees, whose future options need to be protected.

UNDP Human Development Report (1990, pp 61–62)

Compare this wording from the Human Development Report of 1990 to that of SLA as envisaged by DFID:

> The livelihoods approach puts people at the centre of development. People—rather than the resources they use or the governments that serve them—are the priority concern. Adhering to this principle may well translate into providing support to resource management or good governance (for example). But it is the underlying motivation of supporting people's livelihoods that should determine the shape of the support and provide the basis for evaluating its success.Website: The DFID approach to sustainable livelihoods (www.nssd.net/references/SustLiveli/DFIDapproach.htm, accessed September 2009);

There is clearly much overlap between the two and it is easy to see how SLA can almost be a framework for achieving human development, at least at the scale of the household and community. However, the phrase "it is people, not trees, whose future options need to be protected" in the HDR (1990) can be misleading as it may imply that the environment is of secondary importance; that people can be allowed to systematically destroy their environment if it means that they can enhance their livelihood. The concept of human development and indeed a sustainable livelihood certainly does not seek to facilitate livelihood at the expense of the environment:

> However, while it starts with people, it does not compromise on the environment.
> Indeed one of the potential strengths of the livelihoods approach is that it 'mainstreams' the environment within an holistic framework.
> Carney (1998)
> Short-term survival rather than the sustainable management of natural capital (soil, water, genetic diversity) is often the priority of people living in absolute poverty. Yet DFID believes in sustainability. It must therefore work with rural people to help them understand the contribution (positive or negative) that their livelihoods are making to the environment and to promote sustainability as a long-term objective.
> Indicators of sustainability will therefore be required.
> Carney (1998)

It is sometimes said that human development as encouraged by UNDP has more in common with the earlier 'basic needs' approaches to poverty measurement and alleviation than to Sen's vision of capabilities (Srinivasan 1994; Ravallion 1997). 'Basic needs' is a generic term which covers approaches based on the notion that human beings need a basic set of resources (food, water, clothing, shelter etc.) to survive. Exactly what these are can vary depending upon who is defining 'basic needs'. Sen does make a clear distinction between 'basic needs' and capabilities (Sen 1984, pp. 513–515), but even so the influence of human development on SLA is clear (de Haan 2005).

Another influence on the notion of sustainable livelihood and indeed SLA is the field of 'new household economics' which grew during the 1980s and its focus on household labour, income generation and expenditure, even if there were recognized limitations to seeing households in such mechanical terms:

> The major shortcoming of structural–functional and economic approaches to the household is the neglect of the role of ideology. The socially specific units that approximate 'households' are best typified not merely as clusters of task-oriented activities that are organized in variable ways, not merely as places to live/eat/work/reproduce, but as sources of identity and social markers. They are located in structures of cultural meaning and differential power. Guyer and Peters (1987, p. 209). Cited in de Haan (2005, p. 3)

Numerous publications in the 1980s sought to understand households in the developing world, especially in agrarian societies in Africa. A flavour of this is found in the writing of Jane Guyer who did much of her research in Nigeria (Guyer 1981, 1992, 1996, 1997).

Indeed there are so many influences which have helped spawn SLA that it is helpful to set them out as a chronology. Table 2.1 is based upon such a chronology originally set out by Solesbury (2003) covering the period 1984–2002 and which has been expanded, to include some other influences that may well have been important.

Indeed given this long history it can reasonably be asked what exactly is new about SLA? The focus on households and participation is not new and neither is the attempt to understand and integrate aspects considered important for development. Even the 'sustainable' in the name of SLA has a long heritage, and the same applies to the idea of making interventions (including policy) evidence-based. The reader may understandably consider that SLA is nothing more than a new(ish)

Table 2.1 Sustainable livelihoods chronology (after Solesbury 2003, pp 3–4)

1960s/1970s	Integrated Rural Development projects funded by the World Bank and others
	Concept of Agro-ecosystem Analysis emerges (combines ecological, social and economic components)
	Gradual evolution of 'systems' approaches such as Farming Systems Research and participatory methods in development (RRA and PRA) through the 1970s and into the 1980s
1980s	1980s sees the rise of New Household Economics
1984	Long refers to 'livelihood strategies' in his book 'Family and work in rural societies' (Long 1984)
1985	Amartya Sen's book *Commodities and Capabilities* is published by Oxford University Press
1987	The World Commission on Environment and Development (WCED) publishes its report: *Our Common Future* (often referred to as the 'Brundtland Commission report'). The notion of 'sustainable livelihood' is referred to
1988	International Institute for Environment and Development (IIED) publishes papers from its 1987 conference: *The Greening of Aid: Sustainable Livelihoods in Practice* (Conroy and Litvinoff 1988)
1990	United Nations Development Programme (UNDP) publishes the first Human Development Report (HDR) which included the Human Development Index (HDI); an amalgam of income, life expectancy and education regarded as important components within capability. The HDR is published each year since 1990 and include updated figures for the HDI and a suite of other indices
1992	United Nations (UN) holds a Conference on Environment and Development; the Earth Summit. Held in Rio de Jeneiro
	Institute for Development Studies (IDS) at the University of Sussex in the UK publishes 'Sustainable Rural Livelihoods: Practical concepts for the 21st century' (Chambers and Conway 1992)
1993	Oxfam starts to employ SLA in formulating overall aims, improving project strategies and staff training
1994	CARE adopts household livelihoods security as a programming framework in its relief and development work
1995	UN holds World Summit for Social Development
	UNDP adopts Employment and Sustainable Livelihoods as one of five priorities in its overall human development mandate, to serve as both a conceptual and programming framework for poverty reduction
	IISD publishes *Adaptive Strategies and Sustainable Livelihoods* (Singh and Kalala 1995), the report of a UNDP-funded programme
	SID launches project on Sustainable Livelihoods and People's Everyday Economics
1996	*Adaptable Livelihoods: coping with food insecurity in the Malian Sahel* (Davies 1996) is published by Macmillan
	DFID invites proposals for major ESCOR research programme on Sustainable Livelihoods
	IISD publishes *Participatory Research for Sustainable Livelihoods: A Guidebook for Field Projects* (Rennie and Singh 1996)
1997	New Labour elected by a landside (179 seat majority)
	New Labour government publishes its first White Paper on international development, *Eliminating World Poverty: A Challenge for the 21st Century*

(continued)

Table 2.1 (continued)

1998	DFID's Natural Resources Department opens a consultation on sustainable livelihoods and establishes a Rural Livelihoods Advisory Group
	Natural Resources Advisers annual conference takes Sustainable Livelihoods as its theme and later publishes contributory papers: *Sustainable Rural Livelihoods*: *What Contribution Can We Make?* (Carney 1998)
	SID publishes *The Sustainable Livelihoods Approach, General Report of the Sustainable Livelihoods Project* 1995–1997 (Amalric 1998)
	UNDP publishes *Policy Analysis and Formulation for Sustainable Livelihoods* (Roe 1998)
	DFID establishes the SL Virtual Resource Centre and the SL Theme Group
	IDS publishes '*Sustainable rural livelihoods*: *a framework for analysis*' (Scoones 1998)
	The FAO/UNDP Informal Working Group on Participatory Approaches and Methods to Support Sustainable Livelihoods and Food Security meets for the first time
1999	DFID creates the Sustainable Livelihoods Support Office and appoints Jane Clark as its Head
	DFID publishes the first *Sustainable Livelihoods Guidance Sheets*. These have been regularly updated and are available at www.nssd.net/references /SustLiveli/DFIDapproach.htm#Guidance
	DFID also publishes *Sustainable Livelihoods and Poverty Elimination* (DFID 1999) and *Livelihoods Approaches Compared* (Carney et al. 1999)
	Presenters at the Natural Resources Advisers' Conference report progress in implementing SL approaches and DFID later publishes these in *Sustainable Livelihoods*: *Lessons from Early Experience* (Ashley and Carney 1999)
	Overseas Development Institute (ODI) publishes 'Sustainable Livelihoods in Practice: early application of concepts in rural areas' (Farrington et al. 1999)
	DFID establishes the Sustainable Livelihoods Resource Group of researchers /consultants
	Amartya Sen's book *Development As Freedom* is published (Sen 1999)
2000	DFID commissions and funds Livelihoods Connect, a website serving as a learning platform for SLA
	United Nations Food and Agriculture Organisation (FAO) organises an Inter-agency Forum on Operationalising Sustainable Livelihoods Approaches, involving DFID, FAO, WFP, UNDP, and International Fund for Agriculture and Development (IFAD)
	DFID publishes *Sustainable Livelihoods—Current thinking and practice* (DFID 2000a); *Sustainable Livelihoods—Building on Strengths* (DFID 2000b); *Achieving Sustainability:Poverty Elimination and the Environment* (DFID 2000c); and more SL *Guidance Sheets*
	The Sustainable Livelihoods Resource Group establishes a subgroup on PIP (Policy, Institutions and Processes)
	IDS publishes 'Analysing Policy for Sustainable Livelihoods' (Shankland 2000), the final report from its ESCOR programme
	Oxfam publishes *Environments and Livelihoods*: *Strategies for Sustainability* (Neefjes 2000)
	Mixing it: *Rural livelihoods and diversity in developing countries* (Ellis 2000) is published
	The UK government publishes its second White Paper, *Eliminating World Poverty*: *Making Globalisation Work for the Poor* (DFID 2000d)

(continued)

Table 2.1 (continued)

2001	Millennium Development Goals established
	New Labour wins election
	DFID commissions research on further development of the SLA framework; practical policy options to support sustainable livelihoods
	Sustainable Livelihoods: Building on the Wealth of the Poor (Helmore and Singh 2001) is published
	DFID organises SLA review meeting of officials, researchers and practitioners
2002	World Summit on Sustainable Development (Earth Summit 2002) takes place in Johannesburg, South Africa. Called Rio +10
2012	World Summit on Sustainable Development takes place in Rio de Jeneiro Called Rio +20

name for what in fact are old ideas and concepts. SLA certainly provides a convenient framework (and hence title) which brings together the various points discussed so far, but it is perhaps most distinctive in its roots within economic concepts of capital.

2.5 Capital in SLA

SLA is an example of the 'multiple capital' approach where sustainability is considered in terms of available capital (natural, human, social, physical and financial) and an examination of the vulnerability context (trends, shocks and stresses) in which these capitals (or assets) exist. The five principal capitals often suggested as important to livelihood are presented as a pentagon in Fig. 2.2.

Some have already been mentioned, and are straightforward. For example the man-made physical capitals of buildings and machinery and the natural (non

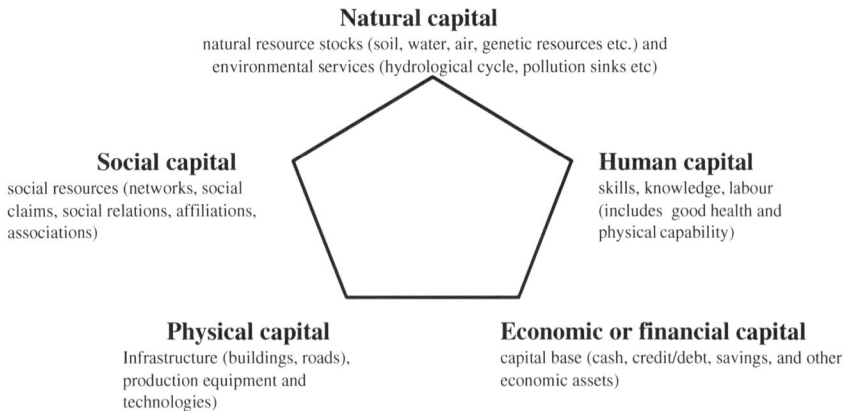

Natural capital
natural resource stocks (soil, water, air, genetic resources etc.) and
environmental services (hydrological cycle, pollution sinks etc)

Social capital
social resources (networks, social
claims, social relations, affiliations,
associations)

Human capital
skills, knowledge, labour
(includes good health and
physical capability)

Physical capital
Infrastructure (buildings, roads),
production equipment and
technologies)

Economic or financial capital
capital base (cash, credit/debt, savings, and other
economic assets)

Fig. 2.2 The five capitals of sustainable livelihood (after Scoones 1998)

man-made) capitals of soil, water, crops and so on. However some are less immediately obvious, such as social networks, knowledge and good health. All are important although clearly the extent of their importance will change from household to household and over time. Indeed people may sacrifice some capital for others if they deem it more appropriate for livelihood, and that switching may reverse at another time (Bebbington 1999). Thus even a relatively simple diagram such as Fig. 2.2 has much embedded complexity and can hide what is in practice a complex dynamic of change in the importance of various capitals for any one household. Even so, attempts have been made to link these livelihood capitals to a measure of poverty; with the assumption being that they provide a multidimensional and inverse proxy for poverty (less capital equates to greater poverty; Erenstein 2011)

The notion of exploring livelihood through such multiple capitals has a long pedigree in economics, but not so overtly within the systems or participatory approaches mentioned earlier. For many the term is limited to describing 'money' (i.e. financial capital held in a bank account or as investments) and therefore the breadth of capitals in Fig. 2.2 may come as something of a surprise. In classical economics 'capital' is a term used to describe a factor of production. Adam Smith (1723 to 1790), the pioneer of political economy, analyzed production by looking at the distribution of costs across the inputs that were required for the process. Money has to be turned into physical inputs before production can occur:

money → payment for capital → production → outputs → revenue

In the classical model, capital underlying such production comprises 'things' such as land or natural resources (minerals, plant products etc.), labour and human-made capital such as machinery. This vision of capital as the basis for production is said to have had its roots within some of the first attempts to record debits (payment) and credits (revenue) within accounting. Note how they are described here as being 'physical' (or tangible). This was indeed the early classical vision of capital—as physical entities that go into production. But this is clearly incomplete as much depends upon 'how' such inputs are used in production; given the right knowledge more can be achieved with less. Therefore since the 1960s economists have taken a broader view and included human capital, such as investment in education and training, within a consideration of production. Indeed O'Neill (2005) has even made the interesting suggestion that ergonomics, the study of the relationship between workers and their environment, can play a significant role in SLA precisely because it seeks to create the conditions that maximise productivity.

It should be noted that what comprises capital within SLA is open to some debate and the five capitals in Fig. 2.2 should not be considered as being definitive, although much can depend upon how broadly the capitals are defined. Serageldin and Steer (1994) suggest that there are four types of capital that need to be considered in sustainability:

- human-made capital (equates to the physical capital in Fig. 2.2)
- natural capital
- human capital
- social capital

and that these need to be expressed in monetary terms—no easy task. Some have also argued for the inclusion of spiritual capital, distinct between social and human capital, which encapsulates the benefits to society provided by spiritual, moral or psychological beliefs and practices. Odero (2006) has also suggested that information should be seen as another capital and this is distinct from what some refer to as 'intellectual' capital. A dilemma is that spiritual and intellectual capital may appear to be subsets of social capital, and indeed the boundaries can be quite fuzzy. It may come down to what is associated with an individual or with society as a whole. Some have also argued that capital within SLA should not just be seen as factors underpinning production, in a mode akin to that adopted by Adam Smith, but in more nuanced ways in terms of how people can engage with others and what such engagement provides for all. For example, there is the following point made by Bebbington (1999, p. 2022):

> People's assets are not merely means through which they make a living: they also give meaning to the person's world.

Capital is therefore a means by which people can "*engage more fruitfully and meaningfully with the world, and most importantly the capability to change the world*". Thus they are not just 'things' that go into a production process but also a basis for power to act and ultimately to bring about change in society. Hence Bebbington (1999) suggests that these capitals take on three distinct roles:

- vehicles for instrumental action (making a living)
- hermeneutic action (making living meaningful)
- emancipatory action (challenging the structures under which one makes a living)

This is a much more nuanced meaning of capital and arguably can embrace information technology and enhanced connectivity via devices such as mobile phones (Sey 2011). It can also embrace culture, religion and recreation as these help to make living meaningful, but such hermeneutic and emancipatory uses of capital are often not included with SLA; a point to be discussed later.

At a most basic level social capital covers the connections between people; or social networks. Its first use within the academic literature was in a paper published in 1916 by L.J. Hanifan. He was researching a rural school community center and explained what he meant by social capital as follows:

> In the use of phrase social capital I make no reference to the usual acceptation of the term capital, except in a figurative sense. I do not refer to real estate, or to personal property or to cold cash but rather to that in life which tends to make these tangible substances count for most in the daily lives of a people, namely, goodwill, fellowship, mutual sympathy and social intercourse among group of individuals and families who make up a social unity…

The latter part of this definition can be summarised as a 'social network' but this is a loose term. Indeed membership of social networks can be quite fluid. Lyons and Snoxell (2005) for example explored changes in social capital of migrant traders to Nairobi in Kenya and showed that while they bring with them and utilise the social networks they already had in place (what the authors term as 'inherited' social capital) they quickly developed new ones in the urban context. Networks were built in an opportunistic fashion but were nonetheless critical to survival. Korf (2004) came to similar conclusions regarding the importance of social networks after using SLA to explore livelihoods in war torn areas of Sri Lanka, especially linkages with key holders of power. Grant (2001) refers to what she calls 'bonding' and 'bridging' social capital, with the former influencing the ability of a group to act together while the latter is the ability of a group to collaborate with others. However, it is important to avoid the simplistic assumption arising out of these studies and many others that being part of a network is always a 'good thing'; even words such as goodwill, fellowship and mutual sympathy used by Hanifen (1916) bring out this sense of the positive. This is understandable. For example, a group can provide support against workplace exploitation or provide better access to resources, but this may not necessary be so and social networks can be a constraint which limits livelihood options. Portes (1998) identifies a number of 'negatives' that can be associated with social capital:

- exclusion of 'outsiders'
- excessive claims on group members such as fees
- restrictions on individual freedom as a result of rules and regulations imposed on group members
- downward leveling norms

An example of the importance of social capital relevant to the Nigerian case study to be discussed later is that associated with faith-based groups. In Christianity all Churches and denominations play a unique role in the provision of social capital. Goodwill, fellowship, mutual sympathy and networking are the hallmarks of these institutions but their roles go far beyond this in the 20th and 21st centuries. Experience in Africa has shown that desirable change can take place within such groupings. Just as a group can provide support and lobby against exploitation in the workplace so too can groups be enabled to examine their own situation and explore the means whereby the members can reach their full potential. To the Christian the 'Glory of God' means that people are more fully alive and where better than in a place of reflection and prayers to ponder on such matters. Most religious groups now concentrate more on a people-centered approach than they did previously when the emphasis was on the group or society and not on the individual. This shows that groups can gradually change and understand the need for human development at a personal level; it is clear to many that lives and livelihoods develop to the extent that people take control of their destinies and understand their rights to a means of living. But poverty is often so great that people cannot think or even feel along such lines so worried are they about how they can provide the next meal or get a sick child to hospital. It is fair to say that Churches have always made great strides in addressing the issues of poverty as it is their specific remit to take care of the downtrodden and the poor. Churches have taken a more radical approach in the past 50 years and it is at grass root level that this enjoys greatest strengths. This often demands being counter cultural but reflective practices such as those introduced by Paulo Freire (1970) and adapted through various teaching methods such as 'Training for Transformation' have and are gradually changing communities. Root causes of problems have been tackled without causing revolutions and gradually women and widows in particular find means of improving and demanding their rights. Such groups may appear low key to outsiders but they are active and are all about the welfare of their people in a society where social welfare does not exist. They organise contingency funds from their meagre resources and are the first and most efficient in dealing with the tragedies especially maternal deaths when strike regularly. Women can often be more constrained in the decision making as men have to give permission for example for a women to be admitted to hospital for a caesarean section. But with more education this is slowly changing. Groups can help facilitate such change. The question the authors often pose is what change would be possible without such groupings?

However, while noting the above contributions it has to be acknowledged that the term social capital has taken on a number of hues and can be argued to hide as much as it reveals. As a result, it has even been argued that *"some authors employ the term not for its conceptual cogency but rather in the hope that it might give their work more visibility."* (Bebbington 2002, p. 1). Indeed Bebbington (2002) argues that one of the problems with social capital may be that it is a label covering too many situations.

Compared with social capital the natural capital component of the SLA is arguably more tangible. Natural capital can comprise goods and services such as the soil for growing crops and trees, water for drinking, washing, cooking etc., uncultivated plants for food and medicine, wild animals for food and so on (Daily 1997; Norberg 1999). Indeed the natural capital and related services it provides has been viewed in a different way within the currently in vogue concept of 'ecosystem goods and services' (EG&S; Costanza et al. 1997; de Groot et al. 2002; Fisher et al. 2009). This was a term first used by Ehrlich and Ehrlich (1981), although, the concept of an ecosystem and the human 'place' within ecosystems and the damage that can be wrought is much older. The difference is that with EG&S the goods and services are allocated a monetary value rather than simply being recorded and evaluated, although these are necessary first steps. The logic behind such economic valuation has been set out in the seminal paper on EG&S by Costanza et al. (1997, p. 253):

> Because ecosystem services are not fully 'captured' in commercial markets or adequately quantified in terms comparable with economic services and manufactured capital, they are often given too little weight in policy decisions.

More recent approaches to EG&S have tended to downplay the need for monetary valuation, but the origins were very much in this sense of providing an equal playing field with other capitals.

The literature on EG&S has expanded rapidly (Fisher et al. 2008a) in recent years and given this wealth of research and literature on the topic it is not possible to go into any detail here. The interested reader is referred to various reports published by the Millennium Ecosystem Assessment (2005a, b) and available at www.maweb.org/en/Index.aspx. The Millennium Ecosystem Assessment (2005b, p. V) categorised ecosystem services into four main types:

> provisioning services such as food, water, timber, and fiber; regulating services that affect climate, floods, disease, wastes, and water quality; cultural services that provide

recreational, aesthetic, and spiritual benefits; and supporting services such as soil formation, photosynthesis, and nutrient cycling.

The inclusion of 'cultural services' in this list is an interesting one given that as
Daskon and Binns (2010) and Tao et al. (2010), amongst others, have pointed out,
SLA often does not adequately address traditional cultural and religious values,
even within the social capital category, and may even see them as a constraint.
In effect the 'services' in the quotation above are of two types. Firstly there are
the processes that take place within ecosystems whether humans are present or not
e.g. soil formation, photosynthesis and nutrient cycling. Humans can alter the rates
of some of these through management but they are natural processes nonetheless
and will take place even if people were not around to influence them. Secondly
there are the services which equate more to 'benefits' that people can gain from
EG&S. Hence recreation is listed as a 'cultural service' but is obviously not a natural ecosystem process. Similarly a forest will produce wood and products that
can potentially be consumed by humans as food and medicine irrespective of
whether humans are present. EG&S can tend to conflate natural processes with the
benefits that people gain from them. Fisher et al. (2008b) suggested that 'services'
within EG&S be subdivided into two types:

1. Intermediate services. These are the natural process that occur within ecosystems, but which can be managed by people to enhance their usefulness to support final services
2. Final services. The benefits which people gain from the intermediate services.
   For example, drinking water, food, medicine and timber.

Even with such subdivision, EG&S is an arguably less clear term than is the
one used in SLA to cover much the same thing; natural capital. At least the latter is a human-centric term placed within a highly human-centric framework that
stresses the human gain from the ecosystem. Maybe EG&S is an over-elaborate
conceptualisation of natural capital?

One of the initial assumptions behind EG&S which does tend to separate
it out from natural capital as envisaged within the SLA framework is that by
providing a monetary valuation of goods and services then there will be substantial resonance with the language that politicians, policy makers and
managers can appreciate, and this will help avoid EG&S being taken for granted
as 'cost free'. All too often the EG&S have not been appreciated until they have
been lost. Hence while an attribute such as biodiversity may be promoted for its
intrinsic value this may not appeal to groups that control the purse strings. But
if loss of biodiversity can be shown to have a monetary cost then there may be a
greater desire to address it (Department for Environment Food and Rural Affairs
2007). But while an economic valuation of EG&S has a certain appeal providing
an economic valuation for its components is certainly a challenge, especially as
they are not typically traded within markets. A local stream provides water for a
range of services but in many places these services are not purchased in monetary terms but are accessed as a common property resource. Also, of course, it is
not inconceivable that such monetary values of EG&S will vary across space and

time. Indeed an ecosystem may remain relatively constant over time in terms of its components and interactions, but change dramatically in terms of the EG&S valued by humans (de Groot et al. 2010). So how is their monetary value estimated? For ecosystem goods, such as fish for example, that have a market price and traded each day then this may be straightforward; these are direct market uses. There may typically be no economic valuation of the ecosystem food chain that supports the fish, but at least the fish have a market price and many of the species are towards the top of their food chain. But there are many other EG&S that are not traded within markets and in these cases economic value has to be determined indirectly via a range of techniques (Farber et al. 2002). One example is the use of a shadow pricing technique such as contingent valuation. Here people may be asked via face-to-face surveys about their willingness to pay for EG&S. The technique has attracted some criticism given that respondents may, for various reasons, over or under-state their willingness to pay (Diamond and Hausman 1994). Another approach that can be applied in some circumstances is to use a travel-cost methodology, where travel distance and frequency are used to construct a demand curve.

There are various 'Payment for Ecosystem Services' (PES) schemes in place by which ecosystem service 'buyers' compensate 'sellers' who agree to protect, enhance, or restore ecosystem services (Engel et al. 2008; Tacconi 2012). One of the classic examples of a PES scheme is that of the Conservation Reserve Program (CRP) in the USA and currently run by the United States Department of Agriculture. The CRP has its roots in the aftermath of the 'Dust Bowl' disaster in the 1930s; an event which influenced the writing of John Steinbeck's classic novel 'Grapes of Wrath'. At present some of the high profile PES schemes are based on carbon markets as a means of controlling greenhouse gas release. Studies exist which have explored the role that PES can play within sustainable rural livelihoods, and problems noted. An example is provided by McLennan and Garvin (2012) for Costa Rica. These authors were critical of the PES in that country as effective mechanisms for linking forest recovery and sustainable rural livelihoods.

As already noted, within the SLA framework it is possible to regard EG&S as a detailed analysis of the natural capital component but there is obviously some spillover into the other components of Fig. 2.2. The nature of such interaction across these capitals, including how culture impacts upon valuation of EG&S and how, in turn, EG&S can help underpin human wellbeing is still said to be poorly understood (Carpenter et al. 2006). As stated by the Millennium Ecosystem Assessment, (2005b, p. 6):

> The degradation of ecosystem services often causes significant harm to human well-being. The information available to assess the consequences of changes in ecosystem services for human well-being is relatively limited. Many ecosystem services have not been monitored, and it is also difficult to estimate the influence of changes in ecosystem services relative to other social, cultural, and economic factors that also affect human well-being.

Perhaps this should not be that surprising as, the EG&S are very much a human construction and interpretation of gain from a complex set of components and processes which exist and evolve in an ecosystem (Boyd and Banzhaf 2005; Costanza et al. 1997; de Groot et al. 2002). Even the 'value' of something as fundamental as biodiversity in maintaining ecosystem services has been questioned by some (Mertz et al. 2007). Clearly there is much still to be discovered about the natural capital component of SLA and a simple quantitative cataloging of the capital that may be available to a community is not enough.

Finally it is worth noting that these capitals interact across space and time and households may reduce or increase some at the expense of others. The clearest example is that financial capital can be used to purchase physical or natural capital and vice versa as physical and natural capitals can be sold. But this interaction between capitals is not limited to the immediate space where people live but can also occur amongst people separated by space. Family members, for example, may live many miles apart in quite different contexts yet they can exchange capital (Meikle et al. 2001). Thus it is necessary to view capitals not in isolation or static but as dynamic.

2.6 Vulnerability and Institutional Context

Once these capitals have been identified and assessed for the contribution they make (or could make) it is necessary to explore the vulnerability context in which they exist; what are the trends (over time and space), shocks and stresses? Shock tends to denote a more sudden pressure on livelihood. For example, a severe flood and drought can seriously affect natural and physical capital in a short period of time. A locust swarm can devastate a crop in a matter of hours. Stress is a term used to denote a longer-term pressure. For example, an economic downturn can take place over years and lead to unemployment and dampened markers for produce and labour. This is admittedly a subjective divide but it does encourage the researcher to consider a range of pressures that could exist. It may be something of a challenge to predict such things although historical trends and modelling can provide clues. The historical legacy could indeed be very important within SLA (Scoones and Wolmer 2003). Clearly it is not only a matter of knowing what is happening now but also what the trends are and will be in the future. Some assets may change little over time (e.g. land and buildings) while others such as cash and social networks can be volatile and depend upon movement of people into and out of the household. For example, increasing population density can result in fragmentation of land holding.

Vulnerability to shocks can also vary. A drought for example will impact upon natural capital and in turn reduce crop yields, but may have little if any effect on other capitals. In the longer term, a severe drought could impact on a wide range of capitals, including social and human as people emigrate. Similarly, flooding may damage physical and natural capital while having little impact on the others. Climate change as a longer-term trend is increasingly being seen as an important factor that can effect such vulnerability for some populations and SLA provides a framework to understand this and how people might adapt (Elasha et al. 2005; Iwasaki et al. 2009; Simon and Leck 2010; Siddiqi 2011; Below et al. 2012). UIy et al. (2011) provide an example of such an SLA employed to explore options for vulnerable communities living in coastal parts of the province of Albay, Philippines. But these authors also make the important point that vulnerability can vary at low scales. Hence capitals will vary in their resilience to different types of shock and the intensity of that shock, and this can vary over relatively small spatial scales; even within a village. Wlokas (2011) came to similar conclusions with regard to the installation of solar water heaters for households in South Africa. However, this was not just a case of variation in local climate as the approach and strategies adopted by the implementers of the solar water heating project could also have an influence on the extent of any benefits on sustainable livelihood that may be seen by households.

Moreover it is necessary to examine the policy and institutional context within which these capitals exist, including the legal context and what 'rights' may, or may not, exist (Ashley et al. 2003). While some capitals may be vulnerable to certain shocks it may be that authorities are able to act and limit any damage which occurs or perhaps provide recompense. While assets may be damaged by flooding there may be publically owned structures in place to reduce the likelihood of the disaster occurring. Similarly, there may be publically funded extension services available which can supplement the knowledge base of farmers or provide advice and help with irrigation systems. It is not only government services which need to be considered here as they may be non-governmental or even private agencies at hand that can provide support for livelihoods. Finally, it is not only a matter of considering each institution in isolation that matters but also the ways in which they do, or do not work together.

The importance of institutions is often reiterated within the sustainable livelihood literature, and in a variety of contexts that go beyond the examples provided above. Institutions influence the natural access to many of the capitals as well as peoples' opportunities and choices. They can help govern social relations and power structures at many scales. Challies and Murray (2011), for example, highlight the importance of institutional support for small-scale raspberry growers in Chile by improving their capacity to comply with safety and quality standards and hence gain and retain market access via the global value chain. Such access to global markets underpins the sustainable livelihood of these growers. Cherni and Hill (2009), in the context of energy supply in Cuba, make the interesting point that the institutional context is a two-way street even if the SLA does tend to focus on households and communities. Thus policies that help the livelihoods of the poor can also help governments achieve their own policy targets. Indeed there are some interesting points which arise when livelihoods are based on undesirable activities. Tefera (2009) provides an example of an SLA applies to growers of khat (*Catha edulis*) in Ethiopia, a crop which is used to produce an amphetamine-like stimulant which is addictive and illegal to either possess or sell in a number of countries and controlled in some others. But the crop does yield a high income (albeit those market prices can fluctuate) and Tefera (2009) points out that a policy of 'criminalizing' khat production and trade is likely to have a negative impact on the livelihood of growers. What is required are alternatives, but the relatively high market price of khat tends to work against a broadening of livelihoods.

Only when vulnerability and institutional contexts have been considered can it be possible to develop strategies that help enhance livelihood (i.e. generate positive livelihood outcomes). The assumption is that these planned outcomes would feedback to enhance livelihood assets and make them more resilient.

2.7 Representation Within SLA

One issue that no doubt would have come to the mind of the reader regarding the above sections is the extent of community involvement required in SLA. As set out above the scale of the SLA has been left rather ambiguous and terms such as household, family, population etc. have been purposely interspersed with descriptions of capitals, resilience and institutions. As will be seen later, the scale of the SLA in relation to the number of people meant to benefit from the insights has varied somewhat in the literature, but this does create a critical question regarding representation. Just who should be included within the SLA to achieve such representation, and equally important how many people should be included? If the intention is to help a community of say 10,000 households then it may simply not be possible to talk to every one of them. But which households should be included and how many of them should be 'sampled' to gain a meaningful insight into livelihoods that would be representative of the 10,000? Would 100 be enough or 1,000? Given that the time, effort and resources involved in exploring the capitals let alone the resilience and institution contexts is substantial then the question over 'representation' is a critical one.

Unfortunately there is no easy answer to the 'representation' question. Access to households can be influenced by many factors and one of these is that a proportion of households may not be willing to take part in the SLA. Hence there may be a tendency towards 'convenience' sampling; including those that can be reached and are willing to take part. Also, the appropriate sample size to provide an adequate (whatever the terms 'appropriate' and adequate' may mean) representation of a population is something of an art form. Indeed this is the question asked more often than not in any social enquiry. Statistical theory can provide some guidance. There is a basic equation in statistics that gives the required sample size once there is some notion as to the variation one might see in a sample. In any population, say of 10,000 households, it is possible to take a series of samples of 100 households and ask them for their land ownership (a physical capital) and covert to hectares. It is also possible to calculate the mean land area for each sample as well as the standard deviation (SD); a measure of the variation in land ownership within each sample. The sample means will be different, across the series of samples and each of them is an estimation of the true (overall) population mean (the mean land ownership calculated for all 10,000 households). The standard deviation for the samples will also vary and will again give estimates of the variation in land ownership for the 10,000 households. The samples are providing windows on the population as a whole, but none may give the true value. It is possible that, by chance, one sample does indeed provide the true population mean and SD but that will not be known to the researcher. The variation between these sample means is called Standard Error (SE). Ideally one wants to have as small SE as possible. If the variation in sample means is very large (high SE) then the degree of confidence one can have as to where the 'real' value for the population rests will be low; the precision is low. The smaller the SE then the more confident it is possible to be about where the real mean may rest; in other words the precision is higher. Obviously the sample size is an important concern here. Larger sample sizes are likely to provide better estimates of the overall mean (SE will be small). The ultimate would be to have a sample size of 10,000 households—everyone in the population! Similarly if the variable being measured, in this case land area, is reasonably uniform across households then it too can help reduce the variation in sample means.

Rather than take lots of samples it can be mathematically (and conveniently!) proven that variation between sample means (the SE) can be estimated by the values of N and SD for one sample:

$$SE = \frac{SD}{\sqrt{N}}$$

This equation 'works' irrespective of the distribution of the variable being assessed; whether it is positively or negatively skewed (as land ownership can be) makes no difference. Thus the higher the value of N and lower the value of the SD for the one sample then the lower the value of SE. The equation can be rearranged as follows:

$$\sqrt{N} = \frac{SD}{SE}$$

The gain from this simple rearrangement is that if the SD and SE can be reasonably estimated (or guessed!) based upon experience or perhaps a preliminary sample then values can be plugged into the equation to provide an estimate of the sample size (N) that may be required. Admittedly this is all a bit 'ball park' as one not only requires an estimate of the SD but also some notion of precision that one is willing to tolerate (represented by the SE). Given the same value for the SD, if less precision is required (SE is high) the sample size can be low whereas if more precision is required (SE is low) then the sample size will need to be larger.

The reader should note that the equation shown above is a relatively simple example of how an adequate sample size can be arrived at, and it is by no means the only such example. There are other more sophisticated formulae and some statistical software packages allow the user to 'plug in' various assumptions about precision etc. and the program will provide a suggested value for N. But the example does provide the reader with a taste as to how statistics can help with this issue.

With stratified sampling where the population is assumed to comprise of different groups then this challenge becomes more complicated. In that case it may be necessary to derive different values of N across the groups, and the SD for the same variable may not necessarily be the same across them. For example, the land ownership variable used above might have a quite different pattern across genders, ages and ethnic groups within a single village. There is also a case for smaller sample sizes which allow for a greater degree of in-depth exploration of livelihood (Crouch and McKenzie 2006). Naturally there is a trade-off here with representation. Hence it may be of no surprise to the reader that sample sizes within the SLA literature have not typically been formally set using such formulae.

In the SLA literature the sample sizes vary a great deal. Wlokas (2011) in her study of solar water heaters in South Africa employed a sample size of over 600 households in low-income areas of Cape Town and Port Elizabeth. Fernandez et al. (2010) employed a sample size of 237 households in four Mayan communities of Campeche, Mexico, when using SLA to explore the effect of income strategies on calorie intake. Nguthi and Niehof (2008) used SLA to research the effects of HIV/AIDS on the livelihood of predominantly banana-farming households in Kenya. They conducted a survey with a sample of 254 farming households stratified into two main groups; 75 that were affected by HIV/AIDS-affected households and 179 that were non-affected by HIV/AIDS. In all cases the main methodology for data collection was a questionnaire-based survey which could allow for relatively large numbers of respondents to be included in a reasonable time frame. The surveys were also supplemented with other means of data collection such as open-ended stories of impact. However, in all cases the sample sizes were much smaller than the communities from which they came although the size of the latter was typically not given. There is also the question of depth to consider. In each of these examples the focus was upon only one facet of livelihood—energy, calorie intake and impact of HIV/AIDS respectively—which allowed some honing of questions to take place. The importance of the foci was, of course, set out for each study. But in 'exploratory' situations where one is beginning with a blank sheet and SLA is being used to establish what the main issues might be, then such honing may not be possible or indeed desirable. This will discussed again later.

2.8 The Attractions and Popularity of SLA

SLA is comprehensive and people-centred in a direct sense (Glavovic 2006a; Chang and Tipple 2009; Hogh-Jensen et al. 2010), and depends upon the involvement of those meant to be helped by change as well as their local knowledge (Mercer and Kelman 2010). Indeed this is both a principled and practical stance as it is difficult to imagine being able to implement an SLA without the involvement of these people. Thus SLA forces an engagement with those meant to be helped by an intervention or policy. It cannot be done from an office. In line with participatory approaches in general this provides opportunities for community-based learning where people can learn from each other as well as from outsiders (Butler and Mazur 2007). As a result SLA builds upon the long history of the participatory movement in development, and techniques and methods homed over years of application in stakeholder participation can also be used within SLA.

SLA also represents an acceptance that multiple-sectors have to be considered i.e. it is holistic (Tao and Wall 2009). As Allison and Ellis (2001) have succinctly put it with regard to an SLA they implemented with a population of fisher folk:

> Its [SLAs] chief point of departure is to avoid undue preoccupation with a particular component of individual or family livelihood strategies, in this instance fishing, to the neglect of other components that make their own demands on the resources available to the household.

Krantz (2001, p. 1) puts this need for holism in even broader terms:

> The concept of Sustainable Livelihood (SL) is an attempt to go beyond the conventional definitions and approaches to poverty eradication. These had been found to be too narrow because they focused only on certain aspects or manifestations of poverty, such as low income, or did not consider other vital aspects of poverty such as vulnerability and social exclusion.

SLA forces this wider perspective through its very design, and is especially relevant in situations where people may have multiple contributions towards their livelihood rather than just a single wage or salary (Tao and Wall 2009). It also forces a consideration of interactions and trade-offs. McLennan and Garvin (2012) for example employed an SLA to explore livelihoods in North-West Costa Rica and showed how intervention was necessary to help mitigate the negative effects of 'locally-felt' trade-offs between conservation on the one hand and use of resources on the other. Such trade-offs are common where people have little choice, and thus SLA can help highlight the issues and explore possible solutions. Indeed this is not just an issue for rural populations and SLA has been employed in urban contexts. (Simon and Leck 2010). Wlokas (2011) speaks of a 'Sustainable Urban Livelihoods Approach' (SULA), although the essence of SULA is the same as that outlined above. SLA builds from this existing knowledge and experience-base rather than taking a new direction.

There is an assumption underlying all this in that change happens and livelihoods are dynamic rather than static. The importance of understanding the history of where people are helps in appreciating why things are the way they are and why people do what they do (Scoones and Wolmer 2003). Intrinsic within this is the nature of decision making and the inevitable trade-offs and conflicts that can occur. The inclusion of such dynamics from the outset as a part of the analytical framework provides SLA with a clear advantage, although in practice the piecing together of historical context may be difficult.

Finally, SLA sets out what the objective of an intervention should be; need for diversification for example as a means of limiting exposure to risk. Once this has been accepted SLA sets out a process by which that 'broad vision' can be gleaned. There are no detailed schematics, blueprints or precise methods that 'must' be used, only a framework. Thus SLA is a flexible approach that can be implemented in many different ways depending upon local context and expertise available for the analysis. It can also be used as a framework for developing indicators to help policy makers and others chart progress towards attainment of sustainable livelihood (Bondad-Reantaso et al. 2009; Bueno 2009; Nha 2009).

Given the benefits of SLA it is to be expected that it featured within the academic literature. Figure 2.3 is a plot of the number of journal papers which have the terms 'Sustainable Livelihood Approach' or 'Sustainable Livelihood Analysis' in their abstract. It is therefore a similar analysis to that presented in Chap. 1 with regard to 'sustainable livelihood' and as in that case it needs to be stressed that this is not an ideal measure of usage but it does nonetheless provide a clue. It should also be noted that these are refereed papers; their content has been checked and approved by other researchers. Mention of SLA began in 1999, which more or less matches the publication of the SLA framework by DFID in 1998, but the number of journal papers is still surprisingly low at one paper per year until 2003. Even so the number of papers published per year has tended to remain significantly below 10 for almost all the years in the graph with the exception of 2009. The figure for 2012 is incomplete as it only covers the first two months of 2012.

Nonetheless the extent of the academic literature on SLA is remarkably small which may surprise the reader given the comments already made in this chapter and the previous one regarding what SLA can help achieve. In fairness it needs to be stressed that the origins of SLA are very much as a practical tool for development intervention rather than as a research tool, although there is no intention here to diminish the potential value of SLA within research (Hogh-Jensen et al. 2009, 2010). Indeed there can be a fine line between research and intervention, and the two are often intertwined in approaches such as action research. But it was not a framework intended to aid pure research *per se*. Indeed many of the papers included in Fig. 2.3 are based upon applied research funded by donors such as DFID. They are also typically site specific, usually focussed on relatively small in areas terms in parts of a single country. As a result much of the SLA literature is not necessarily in the 'refereed' domain picked up by the literature searches

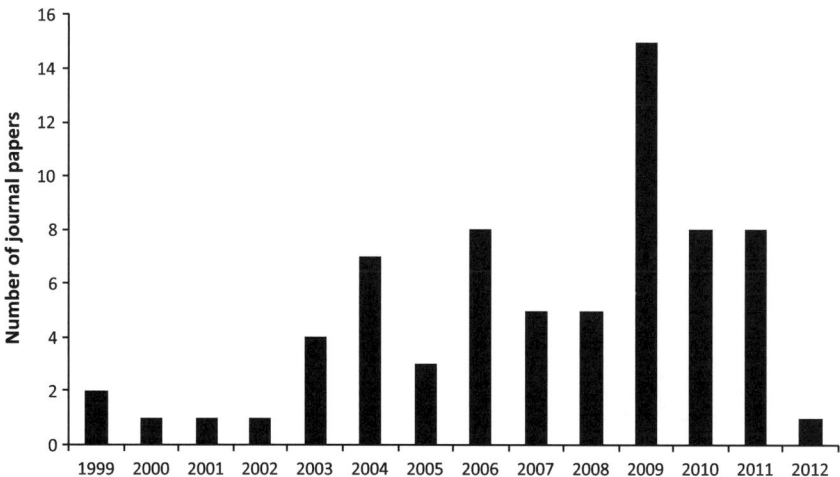

Fig. 2.3 Number of papers that mention SLA in their abstract

used to generate Fig. 2.3. Much of the SLA experience may reside in the so-called 'grey' literatures of project proposals and programme reports required as a condition of receiving funds from development partners. Given that such reports are not always made available to a wider audience then it is to be expected that researchers bemoan the paucity of literature which analyses the operationalisation (rather than theory) of SLA (Allison and Horemans 2006). Even if such reports are understandably site and time specific they can still provide a valuable resource.

SLA is a practical framework for analysing a concept of sustainable livelihood, and it is perhaps no surprise that this wider concept has had much greater reporting within the academic literature; a point made in Chap. 1 with regard to Fig. 1.2. Indeed the contrast between Figs. 2.3 and 1.2 is marked in a number of respects. The start date in Fig. 1.2 is 1989; some 10 years earlier than for Fig. 2.3, and the number of paper published per year is generally much higher. Many of these papers will use the term 'sustainable livelihood' as a concept and in most of them the methodology revolves around other frameworks or specific techniques rather than SLA. Thus, for example, a paper may use the concept of 'sustainable livelihood' to broadly cover what the researchers were exploring but the research may focus only on one aspect of it (one of the capitals or institution perhaps). Indeed even the forerunner of 'sustainable livelihood', namely Integrated Rural Development' has still managed to maintain popularity within the academic literature as shown in Fig. 2.4. The history is a longer one, with papers being published in the early 1970s, with periods of relative stasis in terms of publications per year, but the concept has remained in use right through to 2010.

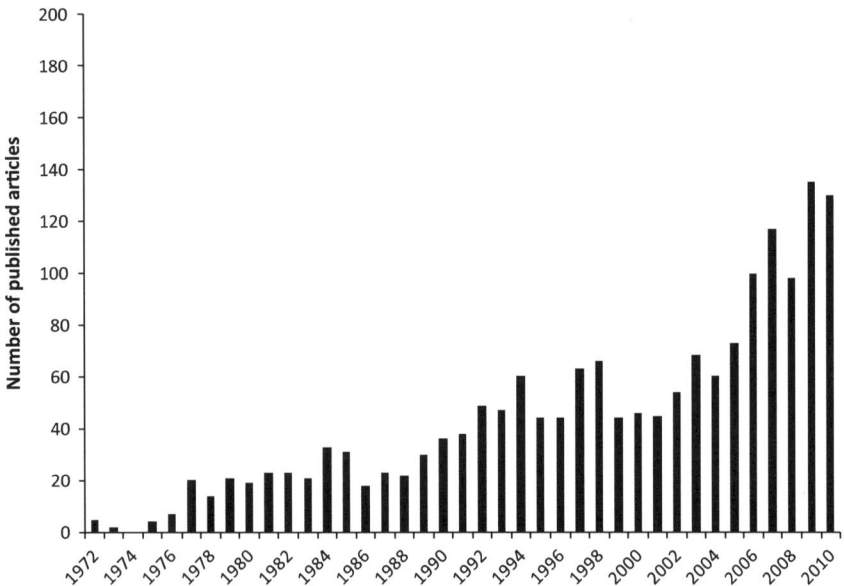

Fig. 2.4 Number of papers that mention 'Integrated Rural Development' in their abstract

One of the lessons of the literature analysis is that the SLA framework is arguably and ironically less popular with academics than is the more abstract notion of a sustainable livelihood. But what are the potential factors that could diminish the value of SLA—at least relative to what one may expect to see given the popularity of the concept of sustainable livelihood? The next section will highlight some of the problems with SLA that have been highlighted in the literature.

2.9   Critiques of SLA

SLA, like evidence-based approaches in general, has had its critics and its proponents are often careful to point out that it is not a panacea (van Dillen 2002; Sillitoe 2004; Toner and Franks 2006; Small 2007; Kelman and Mather 2008). Some of the criticisms are set out as follows, although it has to be noted that a number of these are by no means unique to SLA.

1. For all the people-centred rhetoric of SLA people are strangely invisible in Fig. 2.1. There are capitals, one of which is 'human', influences, institutions, policies etc. but where are the people? The danger is that SLA can become a rather mechanical and quantitative cataloguing exercise which plays neatly into the broad critiques offered by post-modernists and indeed harks back to the 'new household economics' approach and its focus on *"clusters of task-orientated activities"* (Guyer and Peters 1987, p. 209) from which SLA sprang. However, quantification does have advantages; it certainly feeds into the current vogue for numbers and statistics within social policy and thus can have resonance with those using the information to bring about change (Sorrel 2007; Neylan 2008). But SLA has little about 'culture' *per se* even though this is an important consideration for communities (Tao et al. 2010). Indeed if anything 'culture' may be perceived by development practitioners as a constraint to an understanding of opportunities and potential interventions (Daskon and Binns 2010). Also absent from the SLA framework are important considerations such as leisure, and this can have an important impact on resources. For example, in their study of the fishing of Atlantic billfish off the coast of West Africa, Brinson et al. (2009) point out the importance of recreational fishing on the stock and suggest that this should be included in an SLA alongside the more traditional focus on fishing to support livelihood. Hence the paradox of a 'people centred' approach almost entirely avoids some of the key aspects of human existence; people have a culture and also try to enjoy their lives.
2. It is unclear how to analyse and measure capitals within SLA. The pentagon of Fig. 2.2 is a neat representation of important asset groups but each could contain many elements and how are these to be assessed? Is it necessary for all of them to be measured or only some, and if the latter than how is to be determined which to assess? Obviously there is an element of 'context specificity' here, but at least superficially it might seem straightforward. For farming households the obvious physical asset of importance is land and surely land

area can easily be measured? In reality land ownership can be far more complex than this as a household may own many irregular parcels of land which can be spatially scattered at varying distances from the place of residence. Also, of course, there is a difference between ownership of land and access to land through rent or gift. The latter can be volatile and thus constitute a capital in one year but not in the next. Finally there is the issue of 'substitution' of these capitals whereby one could presumably replace another. In economic theory this is the case as capitals are 'production factors' but is it applicable in sustainable livelihoods? Can natural capital really be replaceable by financial capital, and if so how desirable is that for sustainability?

3. Related to point (2) is the importance of trust and openness (Lapeyre 2011). An SLA is reliant upon the participation of those at the centre of the analysis yet the questions being asked, for example asset ownership can be sensitive for all sorts of reasons and it would not be surprising if households withhold information if they felt that the questions are too intrusive (Why is such a question being asked? Will a truthful answer put us in trouble with the law/government?). Again, if land ownership is taken as a seemingly straightforward example, in many countries tax payments are positively related to land area. It would be expected that if a household withheld information about the area of land it owned, it was because it feared that asset could be taxed; the answer would therefore grossly underestimate the asset. This may not be the case, but again it might. The same sensitivity could apply to all asset ownership and potentially distort the outcome of an SLA.

4. An SLA could result in much detailed analysis but how is this to be translated into interventions, policy for example, that will help people? The claim that the process is liberating for participants only holds if those same people have power to bring about change or indeed if they have options. Ahmed et al. (2010) provide an example of prawn fishers in Bangladesh that have limited scope for adjusting their livelihood; restrictions on prawn catches set by government to help the sustainability of the stock has not been enforced precisely because the fishermen lacked alternative livelihood strategies. Indeed power can be a highly skewed property! (Toner and Franks 2006). Some households may be able to adapt to help improve their lot following an SLA while others—frankly—may be able to do little if anything. An SLA, of course, should be able to detect such heterogeneity between households provided the 'sample' size is large enough and the sampling has been designed to pick up such variation. Thus SLA does not avoid the key concern of representation and the 'myth of community' inherent within all participatory methods. The problem is that different actors are involved in the various arrows and neat boxes of Fig. 2.1 and those involved in doing the SLA are not usually the same actors involved in using the information to bring about change, be it through allocation and monitoring of resources or perhaps policy. The danger is that SLAs become an end in themselves and do little more than form the basis for lengthy reports and papers in academic journals. This is by no means an issue solely for SLA, and often voiced in critiques of participatory methods in general (Toner and Franks 2006).

5. While there is an attempt to assess vulnerability (shocks, trends etc.) there is obviously much unpredictability, especially at macro-scales. An historical analysis can help as these can allow some sense as to the likelihood of what could occur even if it does not allow for when. This has become all too clear following the 'credit crunch' of 2008 and its global ramifications, but could also cover more national 'shocks' such as *coup d'etat*, rampant inflation as a result of political instability and even outbreak of disease. Such shocks can have massive impacts at household scales, including abandonment of land and migration and impossible to predict except at relatively short time scales.

6. As a result of the above there is much complexity in SLA. The diagram in Fig. 2.1 may be a neat and simple representation but people's lives are complex. Putting aside the need to consider the wider policy and institutional contexts, and these are complex enough, the first steps of identifying livelihood assets and their vulnerability contexts are 'non-trivial' and there are dangers that arise out of this. As noted by one author, SLA *"belongs to the group of holistic approach that seeks to capture the enormous complexity of development problems, but do so at the cost of focus, depth and analytical clarity"* (van Dillen 2002, p. 251). It certainly can be argued that an SLA exercise has to be based upon inter disciplinarily which in itself is a challenge (Sillitoe 2004) and perhaps goes further than that by evoking trans disciplinarily as new knowledge is *"produced, disseminated and applied in the borderland between research, policy and practice"* (Knutsson 2006, p. 91). If there is to be a 'quick' analysis then the danger is that it could also be 'dirty' driven by the needs of those doing the SLA. The result may be more descriptive (what people do and have) than analytical (why do people have what they have and do what they do?). Ironically the response of the UNDP when designing their attempt to measure capability with the Human Development Index (HDI) focussed on just three elements which they deemed of central importance; income, health care (proxied by life expectancy) and education. Thus human development becomes compressed into just three measures for which data are relatively easily available. In an SLA the information generated may be substantial and decisions have to be made not only about the analyses and interpretation but also presentation to those that need to make use of it.

Some of the above has received remarkably little attention within the SLA literature, which is perhaps surprising given these points are well-known within critical analyses of attempts to use 'evidence-based' approaches to intervention in general (Sanderson 2002; Pawson 2006). Thus while there is an undeniable logic to being aware of the assets available to a household and their vulnerability as a starting point for the framing of a basis for intervention, the creation of this knowledge amongst those employing the intervention is a significant challenge; not simply in terms of a technical issue like measurement but also participation and trust. Any snapshot in time, a catalogue of what assets are present, may be misleading for a variety of reasons. An incomplete 'asset pentagon' may not provide a good foundation and this is before trends in assets are considered. Is there evidence of these increasing, decreasing or remaining the same? Again, some trends may be obvious. For example, land may be divided into smaller parcels as a result of population growth and result in classic indicators of cropping intensity. For others this may be more complex, for example, reliance on memory.

2.10 SLA for Evidence-Based Intervention

The point has now been made through this and the previous chapter, that SLA is typically applied as the basis for intervention; for doing something to help people. In effect it can be a diagnostic 'tool' which provides the evidence-base to help ensure that interventions can be designed to have the most positive impact (Allison and Horemans 2006; Toner and Franks 2006). The intervention itself can take many forms. For example, the intention may be to use the SLA to help design a development project over the short term or perhaps a programme of linked activities over the longer term. Perhaps the SLA points to the need to encourage other sources of income generation or better access to markets. On a larger scale the SLA may be the basis for new policy or changes to existing policy (Kotze 2003; Glavovic 2006a, b; Glavovic and Boonzaier 2007) perhaps by using SLA as the basis for creating indicators (Bueno 2009; Nha 2009) or as a part of methodologies designed to help with decision making (Cherni et al. 2007; Brent and Kruger 2009). In some cases SLA has been used as a tool for evaluation (Mancini et al. 2007). However, while primarily intended as a framework to help guide intervention SLA can also be employed as a 'research paradigm' to help guide the agenda for further research (Hogh-Jensen et al. 2009). The holistic nature of SLA certainly did resonate with policy makers and others and does help to explain the relative popularity it had amongst these groups (Knutsson 2006).

The utilisation of frameworks and tools to provide a solid basis for intervention is certainly not unique to SLA. The literature on ways in which interventions of all types, but especially policy, can be based more on evidence is substantial and goes back many years. In recent times there has been the rise of 'theory of change' as a means to help think through how a project's activities could help bring about change (Funnell and Rogers 2011). In effect this covers the return arrow of Fig. 2.1 which goes from what the SLA suggests should happen to improve matters back to making that change happen. This could involve changes at the scale of the household through to institutions and indeed the state. The latter may involve changes to institutional policy, and if policy is not based on evidence then the chances are that it will not have the desired effect and even have unintended, perhaps negative, outcomes. Gray (2001) for example has set out a gradient of categories that link evidence with policy as follows:

1. Evidence-ignorant policy; policy not even aware of relevant evidence
2. Evidence-aware policy: policy cognisant of but not using evidence

3. Evidence-informed policy; policy considering but not substantially shaped by evidence
4. Evidence-influenced policy; policy changed in some identifiable way by evidence
5. Evidence-led policy; policy that is for the greater part shaped and embedded in evidence about goals and outcomes.

These categories represent a spectrum, from no use of evidence at all in policy (number 1 of the list) to the other extreme (number 5) where evidence leads policy. Between these extremes there are shades of grey, with lots of scope for malleable definitions as to what is meant by words such as 'consideration', 'identifiable', 'shaped' and 'change'. Categories 3 to 5 involve policy being shaped to some extent by evidence, even if identifying this in unambiguous terms might be a challenge. Sorrell (2007) makes the point that this linkage between evidence and policy is supposed to help with a number of problems that one could encounter in category 1 and to some extent category 2. If good quality evidence is not used to inform policy then there is scope for conflict and confusion over key issues as different people bring their own views to bear on the said issues and these may be significantly at odds. Also, there may be an over-reliance on individuals and perhaps *ad hoc* studies that may not necessarily be representative of the wider picture. But there are complex issues involved in the placing of information (even information from an SLA) into a wider context. As Bruckmeier and Tovey (2008) put it, there is a chain linking data (simple facts) to information (where data is interpreted) to what they call knowledge (information placed into context). An SLA may generate information that suggests a clear set of actions, but policy makers place that information within a wider context where those actions may not necessarily be the best options. Indeed Bruckmeier and Tovey (2008) also point out that the chain linking data, information and knowledge can be reversed in the sense that knowledge creates a sense of where the gaps occur and can thereby drive further need for data. This being so, the link between evidence coming from SLA and policy is a two-way street. Policy makers do not only consume information but can also commission its creation. Interestingly the role of those commissioning SLA, and their motivations for doing so, as well as the uses made of any insights, has rarely been discussed within the SLA literature.

Categories 4 and 5 in the spectrum provided by Grey (2001) cover the use of knowledge in what Boswell (2008) refers to as an instrumental mode to distinguish it from other ways in which knowledge can be used within policy (Table 2.2). For example, the fruits of research may not influence policy but can be used as a legitimisation of existing policy decisions (legitimizing knowledge) or perhaps commissioned and interpreted in different ways by contending groups all seeking to influence policy (substantiating knowledge) (Table 2.2).

Notwithstanding these issues, the categories towards the bottom of Grey's list in Table 2.2 have an obvious appeal and have been in vogue within a wide variety of fields for many years; often this may have been more implicit than explicit. Indeed such 'evidence-based policy' has even been described as a

Table 2.2 The use of knowledge

Instrumental knowledge	Legitimizing knowledge	Substantiating knowledge
Organizational structure and substance of research reflect performance targets	Looser fit between structure/ substance of research and policy goals	Structure and substance of research reflect lines of contention
Intensive interest in and take-up of research by decision-makers	Looser ties between decision-makers and research unit	Some exchange between decision makers and research unit
No obvious interest in publicizing knowledge utilization	Clear interest in widely publicizing knowledge utilization	Selected interest in publicizing utilization (to relevant policymakers)

(after Boswell 2008)

modernist-rationalist project. It was especially popular as a mantra with the New Labour government of the 1990s (Sanderson 2002):

> New Labour proclaims the need for evidence-based policy, which we must take to mean that policy initiatives are to be supported by research evidence and that policies introduced on a trial basis are to be evaluated in as rigorous a way as possible. Plewis (2000; cited in Sanderson 2002, p. 4)

New Labour was in power in the UK when SLA began to be adopted and promoted by DFID. Indeed the objectivity implied in evidence-based policy is alluring and as Holt (2008, p. 324) puts it:

> A modern perception of 'evidence-based policy making' is sometimes characterized as a process whereby the 'evidence' is assembled almost independently of the policy options and then through a process of analysis and distillation the policy options and then the preferred policy choice emerge.

This is almost an ideal perception; with evidence providing a range of suggested options and a 'best choice' eventually emerges. An alternative and perhaps less prosaic view of this process is provided by Black (2001) where the evidence is almost "*a retail store in which researchers are busy filling shelves of a shop front with a comprehensive set of all possible relevant studies that a decision maker might some day drop by to purchase*". However Black (2001) does go on to say that the "*the case for evidence based policymaking is difficult to refute.*" However, the problem is that while this may be the case the reality is that policy makers may not necessarily base their decisions on the evidence, including that arising from context. The distinction between information and knowledge made by Bruckmeier and Tovey (2008) has already been mentioned, but the failure of policy makers to adopt recommendations or make use of evidence has been the source of frustration amongst academics and researchers. Huston (2008, p. 1) speaking of the difficulty of making 'evidence-based' approaches a reality makes the following:

> Most social scientists believe that strong evidence should lead policymakers to adopt effective programs and to eschew those that are demonstrably ineffective, but policies sometimes seem to fly in the face of data. The unpredictable and volatile world of social policy has led some researchers to renounce efforts to inform it because they believe that decisions are entirely political and that data are invoked at best only to support a position that someone has already decided to endorse.

The last point in this quotation resonates with some of the types of knowledge set out by Boswell (2008) in Table 2.2. It seeks to remind us that evidence, no matter how good it is, is but one source feeding into a decision-making process. Huston (2008) suggests that in reality decisions can be influenced by the 'four I's':

- Ideology
- Interests
- Information
- Institutional contexts

Only one of these 'Is' is 'information' while the others are much more elusive and indeed in the case of interests and institutions can be quite ephemeral. Black (2001) reinforces this point:

> Clearly, research has only a limited role because governance policies are driven by ide-ology, value judgments, financial stringency, economic theory, political expediency, and intellectual fashion. It would be naive and unrealistic to expect research to provide evidence to clinch arguments about governance policies.

Hence many factors influence decision making besides evidence, and this needs to be considered within any theory of change. Information is just part of the picture. This can be frustrating for those who have dedicated much time and energy to an SLA which would appear to provide clear and workable recommendations to help improve the livelihoods of a community. A classic example of this can be seen with transport policy and this is not a million miles away from an analysis of sustainable livelihood. While in the developing world livelihoods may be more localised to the immediate area where the household resides, although marketing of produce and services can involve significant travel, in the developed world it is not unusual for people to commute long distances to earn a wage. Often this is because paid employ-ment is not available where they live or because the salary is higher elsewhere. Transport policy is an important consideration for their choices, but as Himanen et al. (2004) have pointed out, knowledge of what needs or should be done in trans-port policy can push against what policy makers perceive as being acceptable by the public.

> The present authors have noticed that many experts agree—based on scenarios and mod-elling studies—on the main features of the policy packages necessary for improving sustainability (Banister and Stead 2004). In the case of urban development, these main features include transport policies making car travel less attractive and public trans-port more attractive, and land-use policies to increase urban density and mixed land use (Spiekermann and Wegener 2004). However, these policy packages are not implemented because the public—and therefore policymakers— accept only the last part of the above transport policies: improving public transport. The first part, restricting car travel, is not accepted.

Thus a desire for a more sustainable approach on the part of the transport plan-ners clashes with what the commuter may regard as sustainable; in the latter case defined as being cost effective. In the first case it is the promotion of public trans-port and a lessening of the use of the car while in the latter the car is an important option that should not be underestimated. Hence all the environmental evidence may suggest that car travel needs to be discouraged yet the reality is that this is neither possible nor feasible. Himanen et al. (2004) go on to make the following points:

1. some interventions in the name of sustainability may not make sense in an era of change
2. research information does not seem to have been very influential in the choice of sustainability policy mechanism, especially the choice between standards and market approaches
3. even carefully planned and implemented policy actions may provide disappointing results because of unexpected human behaviour and/or misjudges market responses.
4. ways policies are adopted. Planners reconcile demands for sustainability with other public goals (affordability, equity, acceptability).

Thus despite the evidence being strong that car use should be reduced, people want to use their cars and policy makers end up responding to that demand. As already mentioned, this is relevant to SLA as it highlights the importance of 'whose livelihood' is being improved? Indeed one can argue that an SLA conducted on a sample of commuters in the developed world would arrive at exactly the conclusion that their cars are an important aspect to their livelihood and thus potentially help inform these policy makers that any effort to enhance public transport has to address certain concerns before these people will give up using cars. Much depends upon what evidence is collected and for what reason. Huston (2008) makes this point amongst others with suggestions that there are limitations to any evidence-based policy. After all, the audiences for the evidence can be diverse, and have different backgrounds which can influence their ability and indeed openness to use evidence. The quality of the evidence can also vary, and any implications of evidence for action can be interpreted in various ways. Black (2001) makes a further set of observations as to why evidence-based policy is difficult to achieve in practice:

- Policymakers have goals other than effectiveness (social, financial, strategic development of service, terms and conditions of employees, electoral)
- Research evidence may be dismissed as irrelevant, perhaps because it was derived from a context not regarded as being broadly applicable
- Lack of consensus amongst researchers about the research evidence (complexity of evidence, scientific controversy, different interpretations)
- Other types of competing evidence (personal experience, local information, eminent colleagues' opinions)
- Social environment not conducive to policy change (perhaps because other priorities are regarded as of greater importance or because the public may not accept the changes that evidence suggests should be made)
- Poor quality of knowledge purveyors

The last point is especially interesting. Who are these so-called 'knowledge purveyors' and why are they important?

> These are the people who carry the research evidence into the policymaking forums. In central government, civil servants usually have this crucial role. In the United Kingdom, a high turnover of such staff, lack of experience in a particular field, and high workload militate against good quality advice.
> Black (2001)

The role of such purveyors, indeed the means by which information is communicated is also rarely, if ever, discussed within the context of SLA. If the SLA derives new information that needs to be brought to the attention of those meant to act on it then how is this communication to take place and by whom? It may not necessarily be the case that those implementing the SLA are responsible for this. Maybe the funder ('owner') of the SLA just wants the job done and a report produced. The implementers may wish to publish the results in another outlet, such as an academic journal or perhaps an academic conference, but the objective here may not necessarily be to convey this knowledge to policy makers. The result so often is that the SLA is 'done' to generate this information but the users may be ill defined and, even if they are clearly defined, the means by which this knowledge is conveyed to them may rest on the assumption that they will read the report or journal papers available. This may be wishful thinking.

Empirical evidence for the importance of communication between researchers and policy makers is available. For example, Choi et al. (2005) reviewed a number of studies which explore the influences that help policy makers make use of research and a summary is provided as Table 2.3. Some of these elements have already been mentioned, but others such as personal contact are not perhaps what one would expect. This appears to head both lists but one wonders how often this occurs with SLA. Do those responsible for the implementation of the SLA have an opportunity to communicate the findings directly to those who will ultimately make the decisions over what to do, or is the communication only via reports and policy briefs?

Given all of the above, it is perhaps unsurprising that the 'rationality-modernity' which underlies such 'evidence-based' approaches, has been critiqued from a number of angles most notably from the constructivist/interpretivist position. Here it is argued that the social world is a complex one and there are real dangers is treating it in a way which suggests that it can be deconstructed to derive actions that will lead to a simple cause-effect mechanism. SLA for all of its efforts to accommodate the breadth of a social world will inevitably fall short of a true appreciation of the complexity. Indeed such critics argue that the evidence that forms the basis for evidence-based policy is itself value laden as humans have made prior decisions over what information to collect and how; these may be influenced by their perspectives. SLA is not immune from this; no matter how 'objective' the framework is presented there is still much scope for bias by directing the means by which data are collected and interpreted. With many SLAs, where those funding and implementing the process may even be external to the communities they are investigating, then the potential for misreading is rife and the dangers are

Table 2.3 Facilitators and barriers to use of research by policy makers, identified in a systematic review of 24 interview studies

Facilitators to use of research by policy makers	Number of studies
Personal contact between scientists and policy makers	13
Timeliness and relevance of the research	13
Research that includes a summary with clear recommendations	11
Research that confirms current policy or endorses self interest	6
Good quality research	6
Community pressure or client demand for research	4
Inclusion of effectiveness data	3
Total studies	24
Barriers to use of research by policy makers	
Absence of personal contact between scientists and policy makers	11
Lack of timeliness and relevance of research	9
Mutual mistrust between scientists and policy makers	8
Power and budget struggles	7
Poor quality of research	6
Political instability or high turnover of policy making staff	5
Total studies	24

(tabulation of data provided by Innvaer et al. 2002)

perhaps more readily apparent to the reader. But this issue highlighted by the constructivist/interpretivist critics is not restricted to SLA. For example, social deprivation is a complex concept to define let alone measure, yet in England the Index of Multiple Deprivation (IMD) comprises a relatively small number of components as shown in Table 2.4. The table shows the construction of the IMD over the three years that it was used to assess social deprivation; 2000, 2004 and 2007. The table lists the various 'indicator themes' (each comprising a collection of indicators) employed to measure social deprivation and the relative contribution that the indicators made to the overall index. Shading of the cells is used to show whether a particular 'indicator theme' was included for that IMD. The detail is not important but even a cursory glance at the table shows how the IMD has evolved over a relatively short space of time. In 2000 the IMD had six 'domains' covered by 32 indicators, and half of the index was derived from just two of the domains; income and employment. The 2004 version of the IMD had some overlap with that of 2000 but included two new domains covering crime and the 'living environment' (air quality, houses without central heating, quality of private sector housing stock and traffic accidents). The 2007 version of the IMD was broadly similar to the 2004 IMD but the geographical scale over which it was assessed changed. This shows that over only seven years not only has the vision of IMD changed regarding what was seen as important but also the geographical scales over which it was assessed changed. It should also be remembered that this change has been driven by a combination of an evolution in the ways in which social scientists envisage social deprivation along with availability of quality data. In effect the IMD, for all its intricacy and empiricism of which Table 2.4 can only provide

Table 2.4 Summary of the indices of multiple deprivation (IMD) for England

Domain of social deprivation	Indicator theme used to assess social deprivation (each theme comprises a number of separate indicators)	Index of multiple deprivation		
		2000	2004	2007
Income deprivation	People in income support			
	People in income based job seekers allowance			
	People in family credit /working families tax Credit			
	People in pension credit			
	People in child tax credit			
	People in disability working allowance/disabled persons tax credit/benefits			
	National asylum support service supported asylum seekers			
Employment deprivation	Unemployment (claimants, jobseekers allowance)			
	People on new deal options			
	Incapacity benefit recipients			
	Severe disablement allowance claimants/recipients			

(continued)

Table 2.4 (continued)

Domain of social deprivation	Indicator theme used to assess social deprivation (each theme comprises a number of separate indicators)	Index of multiple deprivation		
		2000	2004	2007
Health deprivation and disability	Mortality ratios for men and women at ages under 65 Years of potential life lost			
	Acute morbidity			
	Limiting long-term illness			
	Proportion of births of low weight ($< 2,500$ g)			
	Emergency admissions to hospital			
	Disability rates			
	Adults under 60 suffering from mood or anxiety disorders			
Education, skills and training	Working age adults with no or low qualifications			
	Children aged 16 and over who are not in full-time education (school or school level)			
	Proportions of those aged under 21 not successfully applied for/entered higher education			
	School performance data (Key Stages)			
	Primary school children with English as an additional language			
	School absence rate (primary, secondary)			

(continued)

Table 2.4 (continued)

Domain of social deprivation		Indicator theme used to assess social deprivation (each theme comprises a number of separate indicators)	Index of multiple deprivation		
			2000	2004	2007
Barriers to housing and services	Housing	Homelessness Household overcrowding Poor private sector housing Difficulty of Access to owner-occupation	■	■	■
	Access to services	Access to a post office (general post office counters) Access to food shops Access to a General medical practitioner Access to a primary school		■	■
Crime		Burglary Theft Criminal damage Violence			
Living environment deprivation		Social and private housing in poor condition Houses without central heating Air quality Road traffic accidents involving injury to pedestrians and cyclists			

Shaded cell means that the indicator theme was included in the IMD of that year a flavour, is a human construct based on thinking at that time. Indeed for the sake of completeness it is worth noting that the IMD is by no means the only attempt to measure deprivation. For example, the Townsend Index of Deprivation (TID) is an earlier (late 1980s) example but one that is much simpler than IMD (Townsend et al. 1988) comprised of just four components:

1. Unemployment as a percentage of those aged 16 and over who are economically active
2. No car ownership (% of all households)
3. No home ownership (% of all households)
4. Household overcrowding.

The choice of these four variables in the TID is in part influenced by availability of the required data via the UK census. Thus its creator has compromised between what social deprivation is and what data are available to measure it. The creators of the IMD have gone beyond the limitations of what data are available via the UK census but have still had to consider what data may be collected at a cost deemed to be reasonable. In the TID the four variables are combined (with equal weight in terms of perceived importance) to form an overall score. As with the IMD, the higher the TID the more deprived and disadvantaged an area is thought to be.

While it might look quite different to Fig. 2.1 the domains and 'indicator themes' of Table 2.4 can be mapped onto the SLA framework. The income and employment (or unemployment in the case of the TID) domains have a fairly obvious match to sustainable livelihood and have been mentioned in Chap. 1, and the health and education domains are linked to human capital. Access to housing and services span SLA elements such as physical assets as well as supporting institutions. Crime is not mentioned in the SLA, although for some it does, of course, provide an income. It is clearly not a capital but it can negatively impact a number of the capitals and work against social capital as it can diminish trust. The living environment domain also impinges upon a number of the capitals. There are elements here that impact upon human capital in terms of health. Indeed the overlap should not be all that surprising as both the IMD and SLA are coming at the same system from a different angle.

It should be noted that the IMD was commissioned by the government with the intention that it help inform policy; it was not established as an academic exercise. While each of the components of the IMD in Table 2.4 can certainly be justified as being both relevant and important, and can be mapped onto the SLA framework, they are not the only means by which such components could have been selected. Indeed many of the adjectives in Table 2.4 such as 'poor' can be defined in many ways. Given this, the reader can come up with suggestions as to what could be included or omitted in this list of IMD indicators. The relative weighting of the domains (not shown in Table 2.4) is also a matter of opinion or choice.

Both the IMD and the earlier TID attempt to encapsulate the complexity of social deprivation into numerical scores. Thus it is possible to present social deprivation as a league table ranking of different regions of the UK (to help identify 'hotspots' of deprivation) or perhaps present social deprivation alongside other 'measures' such as allocation of resources to help address it. Such indicators and indices (an index combines a number of indicators) have proven to be popular ways of encapsulating complex ideas so that they may easily be digested by policy makers. In SLA it is also possible to use indicators to build up a picture of the various components and interactions. Indicators have a veneer of objectivity but the subjectivity that underpins them can easily be forgotten. Indeed as Turnhout et al. (2007, p. 218) point out:

> However, the ideal of value free science is still very dominant, for example when you look at what is expected of science and at how divisions of labor between science and policy are organized. Science is supposed to produce facts and policy then makes value-laden decisions.

Yet apparent objectivity and 'facts' can be misleading. The IMD is founded upon a degree of subjectivity, or perhaps more accurately of informed opinion. It is a human construct designed to represent in as simple a way as possible an aspect of human existence for those who are meant to help do something about it. But that representation can be highly diverse depending upon who is doing the representing.

However, the counter argument to such a critique of attempts to dissect and model society is that it ultimately can lead to a decision to do nothing. Such post-modernist stances only highlight the complexities of the social world and hence the need for some guidance for human action otherwise it is a recipe for complete abstention from any attempt at intervention, including policy (Sanderson 2002). People are suffering from deprivation so something needs to be done and if this means that simplifications have to be made such as those that underpin the IMD then so be it. It is impossible to represent all aspects of social deprivation that effect people so the choice is either to simplify or—in effect—to do nothing at all—just wring their hands in despair. The IMD may be imperfect but it is better than nothing. Similarly, the SLA may not be able to identify every aspect of people's livelihood and arrive at a perfect set of interventions but at least it is better than having no idea at all.

But nonetheless such critiques do serve to remind us that we cannot avoid the complexity of social systems as a major problem in deriving evidence that can form the basis for interventions (Tavakoli et al. 2000). Given this uncertainty it is possible that even if an intervention is based on evidence then it may not succeed in its intentions or, at worst, perhaps have unintended and negative impacts.

One of the 'I' set out by Huston (2008) for the factors which influence policy is 'institutional context'. This can convey a host of different aspects, including the perceived need for an institution to sustain itself even if evidence suggests that it is no longer needed. Indeed institutions often like to portray themselves as using evidence-based approaches. As Boswell (2008) has pointed out this provides a legitimizing effect, although this is a relatively under-explored field.

> Moreover, contributions in organizational sociology have shown how organizations derive legitimacy through signalling their commitment to knowledge utilization.......
> However, there has been little attempt to develop a theory setting out the conditions under which symbolic knowledge utilization may be expected to occur, or testing these claims through empirical enquiry. This lacuna seems to be especially regrettable for studies of European Union (EU) policy making. It has been argued that EU policy is predominantly

regulatory and technocratic, and that its civil service appears to derive legitimacy from its expertise...... But the absence of a more rigorous theory of the symbolic functions of knowledge has made it difficult to elaborate or test these claims systematically.
Boswell (2008, p. 472)

Thus it can be seen as being positive for an institution to portray itself as being 'evidence based' in the way that it works and so becomes its badge of honour. The alternative is to say that the institution is not or only partially evidence based which can generate a negative image. Those funding SLAs have the same motive to promote it as a means for generating evidence as the basis for action, and may help explain why the approach has been so popular amongst partners providing development funds. However, it is sobering to note that Knutsson (2006) with his work on knowledge integration within SLA and how this has been handled in a number of development-focussed case studies comes to the following conclusion:

> Despite the fact that SLA is often described as an approach to societal problems, such as poverty and lack of development in rural areas, the approach has so far primarily been used as a framework for knowledge production. The knowledge produced by the approach is of course intended to be applied in the context of development projects and programmes, but as the results shows, there are less examples of application than examples when SLA is used as a framework for production or dissemination of knowledge. (Knutsson 2006, p. 95)

At the end of this story, it is perhaps sobering to consider that in the eyes of some, DFID, a pioneer and key champion of SLA since the late 1990s, has subsequently become less enthusiastic (Clark and Carney 2009). One of the issues appears to have been a concern that SLA, as a perceived 'project level' tool, cannot feed into the national-scale policy and budgetary changes with which DFID was becoming involved. There is no reason *per se*, why the lessons of SLAs cannot be a part of this more national-scale change but the perception within DFID appears to have been that the approach is 'small scale' in nature and presumably cannot make much of a contribution towards the evidence that may be required. This is a stark contrast between SLA seen as a framework that generates context specific insights at relatively small scales and tools such as the IMD designed to be applicable at much larger scales. If the focus for change is at the larger scales of the nation state or region then SLA may be seen as having little value. Indeed this highlights one of the problems that often confronts new frameworks and ideas. In the furnace of stark reality, where development and indeed research funding is inevitably limited and the direction can also be driven by political decisions then choices constantly have to be made and it is almost as if there is competition between approaches that could be taken. This sounds odd as surely there is room for all these ideas, and there should be complementarily rather than competition, But SLA is but one approach amongst many that could be taken by a development agency, and the shift in focus (and hence resources) at DFID towards more macro-scale interventions could work against approaches such as SLA deemed to be more appropriate at smaller (project) scales. The policy sands are indeed often in motion and even a framework based upon solid principles and good ideas can lose out in the competition for attention for support. Indeed is that necessarily a bad thing?

In order to address this issue of SLA 'losing out' to competing ideas Clark and Carney (2009) have suggested that the future of SLA within DFID can be secured by the following steps:

1. *"build on concrete achievements and lessons from practice*
2. *develop a simple narrative for livelihoods approaches and link this to other modes of working and DFID's corporate objectives*
3. *review how SLA can be adapted to contribute to current development challenges, including the food crisis, fragile states, economic growth and making markets work for the poor*
4. *address perceived weaknesses of SLA, such as limited analysis of policy processes, ecological sustainability, gender and power relations."*

This is an interesting list of suggestions and will be returned to Chap. 5. The first is at least in part a call for more reporting and analysis of SLA-based case studies as a prelude for such 'building on concrete achievement'. The 2nd and 3rd involve an adaptation of SLA to better meet the perceived agenda of DFID and indeed wider 'development challenges'. The need to evolve SLA to meet DFIDs' "corporate objectives" is an interesting suggestion, but one which would appear to run counter to the founding principles upon which SLA is built—its 'people first' ethos. Why should a framework built on such solid principles have to accommodate itself to the fashions that pertain within development agencies? The suggested adaptation towards "current development challenges" does have more credence, and the challenges listed are no doubt important, but given the all-encompassing nature of SLA then it too does seem somewhat superfluous. As the reader would have seen from the examples provided in the chapter, SLA has already been employed in a variety of contexts and one would have thought that the existing framework could readily address issues of food, markets, economic growth and consider challenges arising between shocks and stresses such as political instability and a "fragile state" environment. Surely that is what SLA does best—the need to consider resilience and the institutional context where livelihoods are being pursued. The fourth point about perceived weaknesses also has an odd 'feel' to it. The point has already been made with regard to the interface between natural capita in SLA and EG&S. Indeed the latter is also an example of a currently successful approach. Hence while it is hard to see why SLA should be perceived as downplaying "ecological sustainability" in fairness it is easy to see how this may be so when compared with the clear ecological focus of EG&S. The perceived weakness of SLA in terms of gender and power relations is far less easy to appreciate given that these should permeate not just the capitals but also a consideration of their resilience and the institutional backdrop. Surely any consideration of the capitals should include factors such as access and control, and it is hard to see why gender and power relations can in any way be minimised within that. Certainly in the case study presented in this book they were major concerns and readily emerged, along with others such as ethnicity and age, as an important fabric. A particular SLA might underplay their relevance and impacts but that be said to be a fault of the framework; more of implementation.

2.11 Conclusion

This chapter has covered much territory; from the origins and form of SLA through to the issues that surround it. Much emphasis has been placed on the practical aspects of SLA and in particular how it is meant to help bring about a positive intervention. The authors do not apologise for this. The mechanics of an SLA are certainly important, and will be returned to in the following chapters; there is undoubtedly no intention to diminish the exploratory value of SLA. Worthy of note is that while the SLA provides a very logical framework for analysis there is much potential variation as to how best it should be done, especially given that compromises are almost inevitably. The resulting trade-offs may mean, for example, that representativeness of the findings is open to question. Of course it is important that an SLA be done 'properly' as this ensures confidence in the knowledge that has been generated. A poorly implemented SLA will be open to much criticism, undermining its findings and conclusions. Similarly the authors are not claiming that publication in refereed journals and conferences of studies that use SLA as part of the methodology is of no value. Such studies provide a wealth of experience to help guide the 'doing' of SLA. But it is important to remember that an SLA is typically being done as a means of informing an intervention and that intending participants benefit. This makes an appreciation of desired change, and how best to bring it about, a vital consideration; just as important as making sure that the capitals, resilience etc. are understood. Part of this will be the communication of the findings to those meant to use them.

Finally, the scale of SLA is an interesting point especially with DFID's recent move towards more national scale interventions and an apparent lessening of the importance of SLA within its agenda. While SLA can be thought of as a means of understanding livelihoods with a broad applicability, it can also be highly context specific. The case for SLA being a basis for a small-scale localised project can make a lot of sense, even if the practicalities of 'doing' the SLA are challenging. This point will be addressed in the following two chapters which explore the use of an SLA by an agency in Nigeria. However, while much is often made of the usefulness of SLA to help inform larger scale (e.g. national) interventions this does have significant challenges. The challenges involved in making decisions more 'evidence-based' are well established, and evidence gleaned from SLA is no different in that regard. This can even result in competition for attention and resource amongst approaches and outlooks, with the 'selection pressure' being the result of a range of factors. While this may sound odd given that the approaches and ideas are typically complimentary in the sense that they operate at different scales, the almost inevitable shortage of resource combined with changing political stances can result in foci which shift around the landscape of potential interventions. For all its holistic and people-centred foundations SLA is no different in that regard and the lessons of what happened within DFID are certainly salutary, as indeed are some of the suggested 'cures'. The oddity here is that a broad-based and holistic framework that can be applied to the livelihood of human being ends up being downplayed because it is not relevant to larger scales of intervention. This point will be returned to in the final chapter of this book.

Chapter 3
Context of the Sustainable Livelihood Approach

3.1 Introduction

All development is undertaken within a wider socio-economic and cultural context, the boundaries of which extend beyond the immediate locality of the project. Cognisance has to be taken of this; otherwise good ideas can end up as bad experiences, damaging the very fabric of the community and society of the beneficiaries. With this in mind, we will provide a brief overview on the socio-economic, historical, religious and political situation in Nigeria, the national context for the case study, with special reference to Igalaland in Kogi State where the SLA was undertaken. The aim is to provide the reader with some of the context for the SLA (including historical context), but in particular to help set out why the organisation facilitating and leading the process felt that the SLA was required.

But first a brief reminder as to the three different ways in which an SLA can be perceived (Farrington 2001):

1. As a set of principles guiding development interventions (whether community-led or otherwise).
2. As an analytical framework to help understand what 'is' and what can be done.
3. As an overall developmental objective in which case this development is the improvement of livelihood sustainability.

The first and second items on this list are strategic, meaning that SLA assists in framing thinking and subsequent action. Here SLA helps to examine what interventions should seek to achieve. However, the second item is a practical one. This is where the SLA framework is used in an analytical (tactical) sense to set out a more precise set of interventions based upon the evidence of what is known to be present. In this case, the SLA becomes much more than the provision of a set of principles to help guide intervention and moves into the realm of action based upon knowledge. But while SLA sets out a framework for what needs to be looked for, the details as to how to do the looking are not provided in a generalisable sense. There is really no such thing as an SLA methodology in much the

S. Morse and N. McNamara, *Sustainable Livelihood Approach*,
DOI: 10.1007/978-94-007-6268-8_3,
© Springer Science+Business Media Dordrecht 2013

same way as there is no single methodology for stakeholder participation. Rather
sensibly, the precise methods for exploring the components and interactions within
sustainable livelihoods are left for individuals to create—based upon local circumstances and their own experiences; one size does not fit all. But this then raises a
number of issues, including:

- Who does the SLA and why? For example, is the implementer external to those
  being 'analysed' or a member of the community? What experience does the
  implementer have of SLA and indeed the context of those being 'analysed'?
- What resources are available to implement the SLA? This includes aspects such
  as personnel, transport and administrative facilities; time is a critical factor.
- What methods should be employed to collect the required information? A wide
  range of options are available here, many of which are mainstream methods in
  the social sciences (observation, interviews, mapping etc.) as well as participatory methods.
- How is the information to be brought together to create the picture of
  sustainable livelihoods? This is perhaps not as straightforward as it may seem.
- Who is the intended 'recipient' of the analysis? This is likely to be those who
  are meant to use the results to bring about desirable change, but there are
  other possibilities. For example, the SLA may be a prelude to a phase of more
  detailed research and development.
- How is the new information to be used to help bring about a desired change?

    These questions are certainly not new within intentional development. They are
the kind of questions asked at the beginning of any development project some of
which have resulted in extensive debates abound the very nature of what is meant
by development and the role of agents external to communities in helping to bring
it about. In many cases those 'doing' the SLA are not members of the community
meant to benefit, and it is they who derive the answers to many of the questions
raised above. Indeed despite the inclusiveness of the language SLA is not a step
change as such within intentional development, and still embodies many of the
issues that rest at the heart of the 'developed-developing' axis and all the issues
of power etc. associated with that. The SLA does not remove those involved from
such a polarity. Instead SLA is an arguably 'better' way of doing what intentional
development has always tried to do, by endeavouring to ensure that interventions
are better planned to make a real difference.
    This chapter and the next will explore some of these issues surrounding the
practice of SLA. It will do this by focusing on one specific case study in order
to illustrate the range of answers to the questions set out above. The case study
is located in Igalaland, Kogi State, Nigeria, West Africa, and has been selected
for a number of reasons. Firstly the SLA took place within a context of long
engagement between a local faith-based development organisation, the Diocesan
Development Services (DDS), and the people of Igalaland. The SLA was therefore
part of a much bigger picture, as they so often are, and one that was (and still is)
evolving over time. Secondly it provides an illustration of an interesting, and to
the authors' knowledge, unique set of motivations on the part of DDS as to why

they felt an SLA would help them in being more effective in making development more enduring and surviving when external help was no longer possible. Thirdly it provides an interesting set of trade-offs that occurred with the SLA. However, for the reader to gain an appreciation of all this it will first be necessary to provide the broader context of Nigeria, Igalaland and DDS. Once this has been covered the decisions surrounding the SLA will be more apparent to the reader. The following chapter will then explore the practical implementation of the SLA by DDS and what they felt they gained from it.

3.2 Governing an African Giant

Nigeria is Africa's largest democracy and in 2010 was the third largest economy (in terms of GDP) on the continent; after South Africa and Egypt. Once the self-acclaimed giant of Africa, it merits the title on the basis of population alone (the population estimated as per the last official census in 2006 is 140 million), and that does not include the extensive Diaspora of Nigerians within and outside of Africa. Indeed it is often said that one out of every five sub-Saharan Africans is Nigerian. Nigeria became independent from Britain in October 1960. At the time of independence many Nigerian and foreign politicians believed Nigeria would become one of the wealthiest and most powerful states in Africa if it pursued the appropriate political, economic and social policies with even moderate success (Falola and Heaton 2008). The country covering 923,768,640 square kilometres is potentially powerful, endowed as it is with a great variety of natural resources: petroleum oil, gas and coal deposits, iron and other minerals, many large rivers and well—established light and heavy industries. Nigeria has been estimated to have the biggest liquid gas reserves in sub-Saharan Africa. Until the political developments in South Africa which led to the abolition of apartheid, Nigeria was politically and militarily the most influential sub–Saharan African country, with oil alone more than adequate to finance desirable developments throughout the country. But instead the history of Nigeria since independence has been one of political, social and economic turbulence, which continues to this day (Falola and Heaton 2008) and which has to provide the immediate backdrop for any attempt to understand livelihoods in the country. There is a myriad of problems hampering progress and the most pressing of these include a paralyzed federal government, the hostilities in the Niger Delta region and the constant threats that have emanated from a series of ethno-religious fault lines. But to detail all this would require a book in itself and the interested reader is referred to the excellent text on the history of Nigeria by Falola and Heaton (2008). Here it is only possible to provide a few examples linked to the history of the country.

As a starting point it has to be noted that the advent of the trans-Atlantic slave trade in the 17th century had a devastating effect on the people of Nigeria, some of which still resonates to this day, but is beyond the scope of this book. It is certainly not the intention of the authors to diminish these impacts, but the story will begin with the early 19th century which saw the development of mercantile capitalism, coupled with the Industrial Revolution in Europe. This created a set of conditions wherein the emerging European industrial powers stood to gain more from the legitimate trade exchange than from the continuation of the slave trade and the latter rapidly died. Efforts were made to intensify trade in agricultural products for the growing European industries and in turn this growth of British commercial interests in Nigeria led to the annexation of Lagos by the British in 1862. In January 1900, Britain formally assumed a protectorate over northern and southern Nigeria; in effect two separate territories each managed separately. The North was predominantly occupied by a diverse group of Hausa-speaking peoples, who were mostly Islamic, while the South was home to a variety of ethnic groups but especially the Yoruba and Igbo. The Southern Protectorate was mostly Christian. All three were later combined into one country. In 1939, Nigeria was sub-divided again, but this time into three administrative units: the Northern, Eastern and Western Region. These divisions still mirrored ethnic boundaries, at least in broad terms, as the Eastern Region was largely Igbo and the Western Region largely Yoruba. These are somewhat sweeping statements as Nigeria has over 250 ethnic groups, and the Yoruba, Igbo and Hausa-speaking peoples are just the largest in numerical terms. The Igala, to be discussed later, are among the minority ethnic groups residing in the 'middle belt' of the country.

By 1949 the British proposed a vaguely defined formula whereby traditional authorities and the mass of Nigerians were to be involved in the process of constitution making as a prelude to full independence. The post-war drive towards decolonisation has been mentioned in Chap. 2 and Nigeria was a part of that story. However, despite this desire for public participation, nationalistic politics turned out to be the real issue of the future and the British as well as the Nigerians were aware of the proportions of this problem. Throughout the 1950s a series of constitutional conferences took place in London and Lagos in an attempt to arrive at an agreed balance. In the same decade, political parties coalesced along existing regional and hence ethnic lines, in anticipation of Independence. The compromise, decided in 1954 by the British, was a Federal Government coupled with a considerable degree of autonomy for the three regions. Most powers were to rest with the Federal Government and regions were to be allocated restricted powers. In 1956, the Eastern and Western Regions were granted self-government and the Northern Region was granted the same powers in 1959. Elections for an enlarged Federal Legislature took place in December 1959, but none of the major parties (themselves largely organised along ethnic lines) emerged with an overall majority. After independence in October, 1960, this was again the problem as none of the political parties emerged as truly national in scope. Thus the ethno-religious fault lines that one sees in Nigeria today have deep roots in the birth of the country, although it has to be said that many have tried to overcome them and create a shared sense of identity that transcends these divides.

Civilian rule came to a close in 1966 as that year witnessed the demise of the First Republic with two coups in the same year. The Civil War which followed in 1967 saw the Eastern Region pitted against the Western and Northern Regions;

with the East looking to break away and form its own country (Biafra). The war ended in 1970 with the defeat of the East and the page was quickly turned on that tragic experience, with General Gowon's judicious policy of national reconciliation. But military rule continued and General Gowon was overthrown in a military coup in 1975. His successors, General Murtala Mohammed and General Olusegun Obasanjo, paved the way for a return to civilian rule and President Shehu Shagari was elected as the first executive president of the Nigerian Second Republic in October, 1979. A new constitution based on the American model was prepared for the inception of the Second Republic, with a total of 19 states rather than the three Regions. But this was short-lived as the army took over once again in 1983, this time under the leadership of Colonel Buhari who was overthrown twenty months later in a bloodless coup after which General Babangida took over leadership. On taking power, he cited self-reliance as the basis of economic recovery which turned into a penitential period the effects of which are still being experienced. The inauguration of a constituent assembly in May 1989 generated much interest as it was widely believed that the members of the assembly would use it as a forum to create political alignments that would form a basis for further political parties and leaders. Women were well represented in it. The new constitution was to be modelled on that of the 1979 constitution, but debate threatened to flounder on the issue of religion. Muslims demanded the inclusion of Islamic (Shariah) courts, despite the fact that Nigeria has a sizable Christian population and indeed many Nigerians follow traditional beliefs, and this impeded progress to such an extent that debate on the topic was banned. A decision on the presidency was ratified under the constitutional assembly; a civilian president would be elected for a six year term, and was to declare assets and liabilities before taking office. This was to ensure the presidency was not a time for accumulating wealth and property. The constitution was promulgated in May 1989.

Two political parties were created by the military government to contest the election for the Third Republic, the Social Democratic Party (SDP) and the National Republican Convention (NRC). These parties were said to be 'national' but unfortunately were still largely perceived to be broken down along the same geographical (and hence ethnic) lines; with the NRC being viewed as a 'Northern' part and the SDP a 'Southern' one. The Presidential election took place in June 1993 but the result was cancelled. The reasons behind the cancellation were never fully explained to everyone's satisfaction but it is perhaps telling that at the very moment the election was cancelled the SDP was ahead. The military temporarily handed over to a transitional government headed by a civilian, Ernest Shonekan, in August 1993, only to come back in November 1993 with another General (Sani Abacha) as Head of State. During all this time there was intermittent and sometimes violent protests in the West of the country as the leader of the now defunct SDP M. K. O. Abiola, a Yoruba proclaimed himself the rightful Head of State. These protests were brutally suppressed. Indeed the human rights record of the Abacha regime came under serious scrutiny by the international community following the trial and execution of Ken Saro Wiwa and other Ogoni activists in November 1995 over the problems of environmental degradation in the oil-rich south-east of the country. The activists pointed out that these areas were bearing the brunt of the environmental degradation associated with oil extraction and transport, yet were gaining little from the Federal and State governments. The executions led to the expulsion of Nigeria from the Commonwealth in 1995 and to the imposition of sanctions, but Sani Abacha declared he would be President for life. This resulted in a general outcry but the problem was solved by the sudden death of the two main protagonists in the national instability, Abacha and Abiola, within a few weeks of each other in 1998. Abacha was succeeded by Alhaji Abubakar, and the long-promised new constitution was signed into law during his short time as president. Fresh elections were held, this time without the government creating political parties, and Alhaji Abubakar handed over power to the winner of the election—Olusegun Obasanjo—who became President of the Third Republic in May 1999. The period from 1999 to the present has been one of relative political stability in the country, with elections peacefully held in 2003, 2007 and 2011. Since 1999 the national political scene has been dominated by the People's Democratic Party (PDP), but presidents have come from both the North and the South. Obasanjo (from the West) was unable to steer the party in the direction he wanted but was elected for a second term in 2003. Yar Adua (a Northerner) who succeeded Obasanjo in 2007 was known for his integrity but no great change was witnessed as he died before his first term was completed. His successor Goodluck Johnson (a Southerner) inherited much volatility on taking office in May 2011 as Nigeria's rank on the Failed State Index shows the country deteriorating from 19th in 2008 to 14th in 2012 (www.foreignpolicy.com/failedstates2012). In the new millennium, there were violent crises in the North as well as in the Middle Belt, Benue State and the southern oil-rich Niger Delta. Hence while it is fair to say that the political scene has witnessed some welcome stability since 1999 (no coups) the volatility and uncertainty within the country has continued. The emergence of the Boko Haram in 1993 but which has only come into prominence in 2010 has accentuated that volatility and added an element of insecurity not known previously. Boko Haram is now considered a major terrorist group whose bombing activities are affecting Nigerian security and arousing well founded fear of attacks akin to that of Al Qaeda in many other countries including the USA who see this group as a major threat to their security. There have been over 20 bombing attacks in the north of Nigeria in the past two years claiming the lives of over a 1,000 people and injuring many others. In the eyes of many Boko Haram activities are designed to achieve explicit political ends many believing they have accomplices within government circles and the group is said to have criminals among its members. Anger seems to be more directed against the government for not finding those accountable than at Boko Haram itself. Loosely translated it means 'western education is forbidden'. Its efforts to Islamise Nigeria do not represent mainstream Muslim population who are embarrassed at such pronouncements. Muslim and Christian leaders and politicians are endeavouring to make peace with Boko Haram but as yet this has not been a success story. Some attribute the Boko Haram problem to the youth unemployment. An element of truth definitely resides there as it is in such places youth can find a place in which to vent their anger.

The economic situation has not improved and the recent drop in oil prices does not augur well for that economic situation. The insecurity in the North is not helping food security and already there is a decrease in productivity. Research staff in the north are also looking for new opportunities elsewhere to avoid living in such insecurity. Airlines are studying the security situation and are concerned about security of staff and passengers. Business in Abuja could be in jeopardy and indeed the north as a whole. To date Boko Haram terrorist attacks are confined to the north but should they be extended to the Christian south what will happen then? Will there be reprisals and retaliation north and south? Could this lead to a second civil war in the name of religion?

3.3 Economic Development in Nigeria

For many years before and after Independence Nigeria was an exporter of agricultural products. The West of the country was noted for cocoa, coffee and rubber, the North for groundnuts and the East for palm oil. Indeed with the outlawing of the slave trade the British interest in Nigeria shifted to its wealth of agricultural products. But this was to change dramatically with the discovery of petroleum oil in the southern states, especially in the Mid West and South East; a discovery which has proved to be both a boon and a bane for the country. Explorations for petroleum oil began in Nigeria in the 1950 and 1960s. By the mid 1960s it was well established that the country had vast resources of high quality crude oil. Indeed it was often speculated that the civil war (1967–1970) was fought in part because of international interest in oil and the 'carbon' inequality within the country whereby the North had no oil but the South, primarily the South East, did. A successful secession of the South East would have had major ramifications for the economy of the North and West of the country and a new state (Biafra) might not necessarily have felt obliged to honour existing agreements which oil companies had with Nigeria. The North and West won the Civil War and Nigeria was held together, thus allowing the proceeds from oil to be distributed throughout the country. Whether this distribution has been 'fair' is a matter of opinion amongst Nigerians. The oil boom can be said to have begun in earnest in the 1970s as Nigeria ended a recovery phase following the war, and there were immediate benefits, the most notable of which being the Universal Primary Education (UPE) programme in 1976 which made primary education both compulsory and free, as well as the development of new infrastructures such as roads and bridges and an increase in the number of states and local governments. The money has also financed the creation of a new capital city, Abuja, located more or less in the geographical centre of the country. Thus the oil wealth significantly increased the number of salaried jobs and the pay of many of those in employment, both within the public and private sectors.

However, at the same time there has been injudicious expenditure and exploitation by both foreigners and Nigerians alike and as noted above a degree of environmental damage inflicted by some oil companies in the south, particularly in the Port Harcourt and Delta regions. The cost was largely born by the losing side in the Civil War. The environmental damage inflicted by some oil companies in the south, particularly in the Port Harcourt and Delta regions, has been substantial with land being spoilt and water polluted. For those living in these places their livelihoods received a major shock as a result, and it is not difficult to imagine their anger as a consequence. Another, more nation-wide problem, brought about by the discovery of oil has been the neglect of the indigenous agricultural sector as its economic importance dwindled by comparison. In 1964 agriculture accounted for 70 % of the total exports, while in some 20 years later this had dwindled to a mere 3.6 %. In contrast, oil accounted for over 96 % of total exports in 1983 and has remained at that high level ever since. Agricultural exports from Nigeria currently account for less than 1 % of the value of all exports, although there is potential for this to expand (Daramola et al. 2008). In contrast, oil continued to dominate the export scene, accounting for over 90 % of total exports as from the mid-1970s to the present. Thus in any discussion of sustainable livelihood it is impossible to avoid the impact of those key extractives; oil and gas.

There is yet another aspect of the oil economy that has had a significant impact on livelihood. The 'oil economy' of the 1970s and early 1980s kept the value of the local currency, the Naira, high on international markets which helped to increase imports and reduce exports of other non-oil-based commodities. Perhaps bizarrely for a country where the majority of the population live in rural locales and depended upon agriculture, it became far cheaper, and politically more expedient, to import food than to promote local production. Even when the Naira came under pressure on International markets as from the mid-1980s the government tried to intervene to keep its value high. Because of the overvalued Naira, commodity prices fell, and imported rice could even be found on sale in the markets of rice producing areas. Nigerian farmers found it difficult, if not impossible, to get good prices for their produce. Extensification and intensification as a means of improving revenue proved difficult. The problem with Extensification was the labour intensive nature of production, and in the case of intensification the relative lack of a supply and maintenance infrastructure; add to this the fragility of the soils. Farmers concentrated on subsistence, and many left to work in the towns and cities. Production of staple crops fell rapidly during the 1970s, but stabilised in the early 1980s. Initiatives taken to boost agricultural production (for example the importation of machinery, inputs and expertise) were not enough to support the industry. Politicians repeatedly paid lip service to its importance, but National Development Plans (NDPs) throughout the 1970 and 1980s failed to reflect this, with typically only 12 % of the total expenditure going to agriculture. By this time migration from rural areas to the cities, resulted in reduced availability of labour in the rural areas and a surplus (and hence unemployment) in the cities; an increase in the importation of food depressing the prices of local produce. There was a serious crisis in the country. The economic policy orientation during the 1970s left the country ill prepared for the eventual collapse of oil prices in the first half of the 1980s.

To counteract the extravagance of the previous decade, Structural Adjustment Programmes (SAP) was introduced by the Military Government of Babangida in 1986 (Moser et al. 1997). The subsidy on oil was reduced and Nigeria began to look at other industries to generate revenue. The aim of SAP in Nigeria, and indeed in many other countries much earlier, was effectively to alter and restructure the economic consumption and production patterns of the economy; price distortions and heavy dependence on the export of crude oil and import of consumer and producer goods had to be addressed and regulated. It was planned that this would correct the imbalances that had developed between 1977 and 1986 and thus help promote agriculture (Mosley 1992). The following austerity measures were introduced:

- 80 % reduction in the subsidy on petroleum products
- Privatisation of government-owned companies to help reduce wastage
- A cut in subvention to most parastatals, of which there were many in Nigeria
- The imposition of a 30 % duty on all imports so as to help encourage local production
- The intensification of the import licensing scheme to ensure that importers had to provide a full rationale
- A shift in government emphasis towards rural development and self- reliance
- Incentives to promote non-oil exports
- Gradual depreciation of the Naira so that exports became cheaper and imports more expensive. The official rate of exchange changed from five Naira to the US dollar in 1987 to 22 Naira to the US dollar in 1992. The unofficial rate of exchange varied between 80 and 90 Naira to the US dollar by 1995. At the time of writing the exchange rate is Naira 160 to one US dollar. As a result, imports became very expensive and the quantity of imported agricultural produce fell.

But while many may have benefited from SAP there were also many losers. Price inflation rocketed and pushed many companies out of business. There were no new employment opportunities, salaries were not paid and many government agencies, including the police, electricity supply, schools, hospitals and universities, no longer functioned. In Igalaland and in many other parts of Nigeria professionals such as teachers and top civil servants vital to the functioning of the country were forced to retire. Most were the cream of the crop of those recently educated with valuable experience and excellent training and irreplaceable at the time. This was done ostensibly (it was said) in the interest of creating new jobs for those coming out of training but without any consideration for the void it was creating. Many of those who were pushed out would have begun work at 14 or 15 years of age and completed professional training as opportunities arose and often based on merit. Most had young families in need of regular income. All this led to untold hardships as gratuity (lump sum payment at the time of redundancy or retirement) was rarely paid on time and small pensions due to those retired took years to materialise.

SAP had a longer term objective and was not meant to provide an immediate improvement in the welfare of the people; it caused much hardship. It did not reduce unemployment nor help price stability over the short term. It priced education and medical services out of the reach of the average person. By 1995/96 the new middle class (i.e. the public and civil servants) was practically wiped out. The rich had become poor and the poor destitute. This resulted in a surge in crime in the cities, towns and even in the rural areas where traditional cultural sanctions were still strong. Corruption became endemic. Many of these effects still pertain to this day as development plans that followed failed hopelessly to restore a balance that redressed the madness of that Black Economy decade. SAP, a reform therapy imposed by the World Bank and the International Monetary Fund (IMF) was expected to last only two years; this was not the case.

> With the introduction of SAP in the mid-1980s the average citizenry was stretched beyond limits. The attendant inflation made survival a grim struggle. The pangs of poverty crept into the meagre resources of the already economically and materially harassed lower classes. Vanguard. 2nd October, 1996.

However, SAP did generate some benefits. It has already been noted that farmers and traders did gain; although that too was threatened as travel to new markets became more expensive due to the high cost of inferior spare parts for vehicles as well as the hike in fuel cost. Following SAP, agricultural production rose by an estimated 2.5 % in 1987, 4.58 % in 1989 and 4.8 % in 1991. In 1988, the Federal Government published the first ever agricultural policy document for Nigeria aimed at redressing the sectors under-development, streamlining policies across all tiers of government and ensuring policy stability. However, it was apparent that any increase in income through self-reliance was eroded by the high cost of education and medical bills. In the medium to longer-term SAP did help bring about some stability to the Nigerian economy and as oil prices have recovered after the turn of the century the country has once again begun to grow, albeit with a relatively high level of inflation.

Thus any discussion of livelihood within Nigeria has to take account of this turbulent political and economic history; all being inextricably intertwined. But this is a macro-picture; the story of changes on a national scale. How have these played out on smaller scales in the country? The next section will focus specifically on one small part of Nigeria; the Igala Kingdom—An area that geographically sits in between the North, West and East with cultural and linguistic contributions from each. At one time the very boundary between the Northern and Southern Protectorates established by the British ran through the middle of the Igala territory. Indeed its strategic location is such that it found itself on the front line of the Civil War and this was to have tragic consequences.

3.4 A Kingdom Discovered

The Igala kingdom (or Igalaland) nestles to the East of the confluence between the rivers Niger and Benue and covers about 4,900 square miles, and represents approximately 1.5 % of Nigeria's land surface area (Fig. 3.1). At the time of

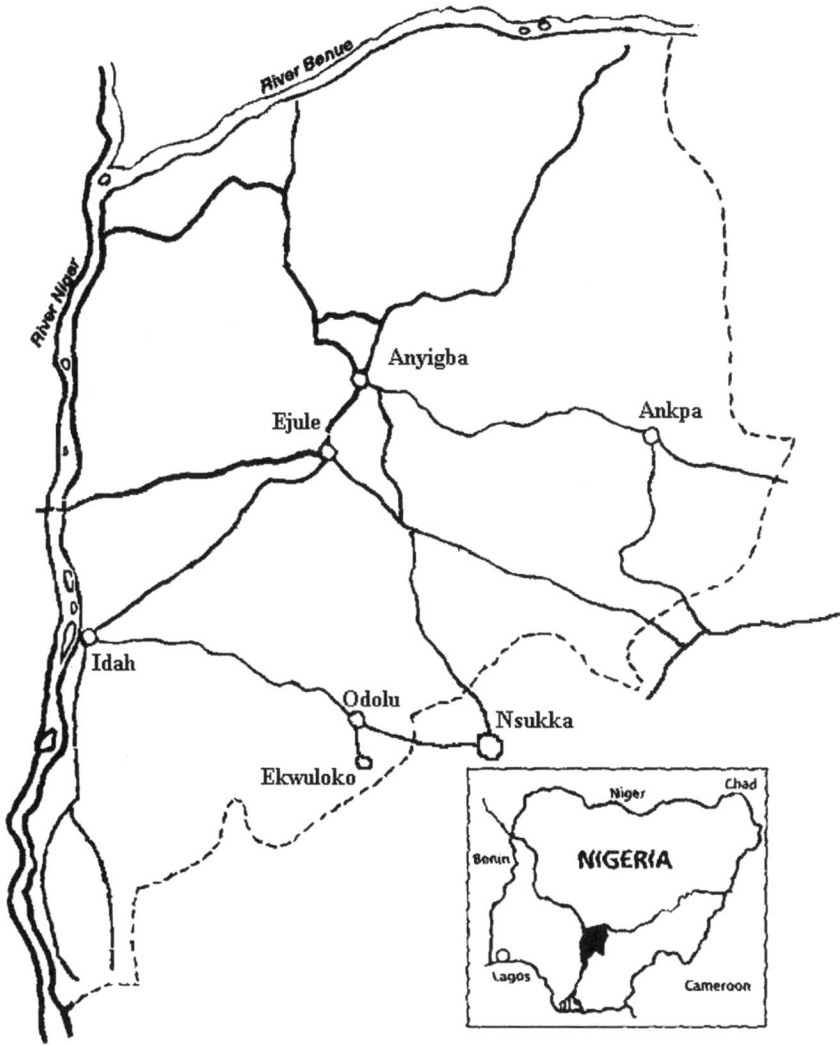

Fig. 3.1 Map of Igalaland showing some of the main towns

writing the population is over one million; comprising a small part of Nigeria's population now estimated to be 150 million. The predominant ethnic group in the area is the Igala, but there are two other smaller groups to the north; the Bassa Komo and Bassa Nge. Religion is mixed, with a high level of religious tolerance as a single family can have followers of the two major religions in Nigeria; Islam and Christianity. Traditional religion is also common as indeed it is throughout Nigeria. So far the area has not been affected by any of the religious disturbances seen in other parts of the country.

The exact origins of the Igala people are not known although linguistic and ethnological evidence suggest they originated from the Yoruba kingdom to the West (Boston 1968; Akinkugbe 1976). Indeed despite the dearth of ethnological studies of the Igala, it is obvious that a great diversity of influences can be recognised in their cultural make up as Igalas differ from locality to locality and therefore their response to a situation or intervention varies widely. The Igala kingdom is among the oldest and more centralised of the kingdoms in the country and was once (see 15th century) a famous and powerful kingdom. Its influence under the traditional King, called the Attah (father), was felt far and wide, extending as it once did to Koton Karifi in the north, to the Nsukka area, in Enugu State in the south, eastward to Idomaland and westwards across the Niger River to the present day Etsako local government area of Edo State. Throughout long periods, the leadership of the Attah was undisputed and he ruled with absolute authority. He was regarded and treated as a super human being, the god of the Igalas. The office of Attah rotated among four royal families, and the Attah of Igala is still the traditional head. Problems and disputes are often referred to him, and he plays a major role in all activities in Igalaland.

The first recorded events in Igala history date back to the end of the 15th century when the Attah of Benin origin ruled from a centre of power and culture at Idah (Okwoli 1973). He established political relationships with other Nigerian kingdoms far and near and it is believed that the son of an Attah of Idah founded the ancient town of Bida, to the North West in the present Niger State. At another period the Nupe kingdom to the North West, also part of the present day Niger State, paid tribute to the Attah of Idah. The importance of the Igala kingdom was pointed out by Allison:

…when the Europeans first entered the Niger, the population along its bank from just North of Onitsha to the neighbourhood of Bara, Niger State admitted themselves tributary to the Attah

This means they paid tax to the Attah, most likely in kind. Allison (1946). Taken from Dawtry (1980).

People from the powerful Benin kingdom (to the west) and the Igala kingdom recognised and respected each other's power and influence. However, the average person may have had only limited mobility within the kingdom and was subjected for the most part to a harsh feudal like existence. Many of the people, especially Igbo and Idoma, who lived on land captured by the Igalas, were exploited and treated as second class citizens. Their traditional religion provided many rules and norms that acted as constraints against what would be detrimental to security, stability and peace in the society. These included limits to any excesses of an over autocratic Attah.

Towards the end of the 15th century the Jukuns began a reign over the Igala kingdom. This war-like people, with expansionist policies, ruled an extensive empire of which Igala was part. But Attah Ayegba refused to be their vassal and eventually expelled them from Igala territory. Not surprisingly he is regarded as the greatest of all the Attahs of Idah. After the death of the Attah Ayegba, the influence of the kingdom declined slowly at first, but more quickly later on. This decline was hastened by the Fulani Jihad (1804) which eventually destroyed the glory of the states of the Niger Benue confluence; Nupe, Igala and Jukuns (Okwoli 1973).

For two centuries (see 18 and 19th) the Igala benefited considerably from the thriving slave trade on the Niger as they received remuneration for the slaves they captured. The environment created by this trade was one of distrust and these fears live on for longer than one might imagine. Oral historians living today still remember the days when women collecting water from the river were accompanied by young boys wielding bows and arrows to protect them. These feelings were possibly inherited either from memories of the slave trade or else a fear of being captured for the ritual killings which were prohibited by law in the 1940s. Today fear is still dominant in the bigger Igala towns largely because of their impersonal nature. Fear and distrust of each other and outsiders continues to dominate life in Igalaland. After the abolition of the slave trade the Igala readily accepted offers of alternative legitimate trade from Europeans (see 1832). In 1841 the British government sent an expedition up the Niger with the objective of making treaties, by whatever means possible, with chiefs along the river, The explorers were insistent that the abolition of slavery be an integral part of such treaties. By 1848 Idah was acknowledged by the Royal Niger Company to be a thriving trading town, the outlet of a huge hinterland to the east. As from the 1850s, peace reigned in Igalaland and trade flourished, with some undulations, for more than a century. The most important trade was the wholesale trade of palm produce. For nearly two hundred years until the late 1960s, ships and barges were loaded with palm produce from Igalaland and beyond at Idah from where they were taken across the world. This trade was lucrative and so its accelerated demise due to the civil war and the later concentration on petroleum products was a real blow to the Igala economy. Although other trade was sought none was as profitable and the people had been made to feel vulnerable.

At the onset of colonial rule (1900), the kingdom was traumatised as it was accidentally divided between two protectorates (Northern and Southern). When the two protectorates were united in 1918, Igalaland was again reunited but this had already created a feeling of instability within what was to become the Federation of Nigeria. Subsequently, as the three Regions of Nigeria came into being, Igalaland became part of the Northern Region. Since the 1960s Igalaland has found itself in various regions and states of the country; Kwara State, Benue State and now Kogi State. Its location between North, West and East meant that it suffered much during the Nigerian Civil War in the late 1960s. The Igalas shared boundaries with Igboland and, because of this, suffered border attacks during the civil war from both the Federal and Biafra armies. In addition, Igbos in Diaspora throughout Nigeria were forced to flee from their homes and businesses to their war torn homeland. The exodus of Igbo traders, tailors, wine tappers, mechanics and carpenters, as well as those in the professional cadre, left a great void in Igala society and economy. The great Onitsha market, the main source of manufactured goods, was within the war zone, so Igalas had little access to the goods they normally used. They in turn lost the main markets for their agricultural produce which were in Igbo towns. The depression that pervaded the scene was tangible and continued for some years after the war ended in January 1970. Igalas did not seem to have the physical, economic, or psychological reserves necessary to sustain themselves. The civil war disrupted life in Nigeria as a whole, but nowhere outside the war zone did the economy disintegrate as badly as in Igalaland and Idomaland, where pre- war Igbo influence was more pronounced than in any other area outside Igboland. Unfortunately the situation continued for some time after the war ended, leaving the people in a state of depression and for some low in self-esteem.

In the last 40–50 years due to the decline in the river trade and other problems such as poor infrastructure, the importance of Igalaland dwindled. The kingdom became virtually isolated from the main areas of economic growth in the country, cut off by the rivers to the north, west and south. The problem was partly recti-fied in 1985 with the completion of the bridge between Itobe and Ajaokuta over the river Niger. This facilitated contact with the Federal Capital (Abuja) and ben-efits the Federation as a whole and not just Igalaland. Although there has been a vast improvement in road infrastructure, many areas still cannot be reached at cer-tain times of the year. As will be noted later, this especially applies to the two vil-lages where the SLA was applied. By and large the general lack of infrastructure has retarded development and progress making life more difficult than in many other places within the Federation. But there are also reasons for this within the Igala themselves. Until the 1950s the Attah and his council resisted many colo-nial changes (including various constitutional developments) welcomed in several other parts of the country, and even to the present day Igalas refuse to take on jobs which they consider menial. Discussions with some Igala elders indicate that this pride can be an obstacle to development. They admit help is needed; mak-ing it clear that for an intervention to be successful tact and patience are required and this coupled with the characteristics found in centralised societies can make work in Igalaland daunting. This is expressed mainly in the male elders' control and tight reins on all activities ensuring strict adherence to tradition. This affects land ownership as well as many other customs that impact upon livelihoods. There is not much incentive to stay at home, and those (especially the youth) with initia-tive and ability find scope outside rather than within Igalaland. Male migration has become the norm and this has impacted upon numerous Igala villages. One finds two extremes, the insular outlook of those who stay at home and the progressive outlook of those who travel beyond its borders.

The smallest unit of settlement in Igala is the hamlet. Hamlets consist of small huts which are clustered in an area, separated from each other by grassland or woodland. In most cases its inhabitants have a blood affinity. The hamlet is some-times called the Oja, but this word can also mean a meeting, a society, or a work group. A village is made up of several hamlets. The offices of the village were changed by the British in the 20th century when the titles of Madaki and Gago were introduced into the now defunct 'Igala Native Authority'. These are Hausa terms from the North of the country and were seen by the British as providing a useful structure for governance. A Gago is the traditional village head, while the *Madaki* is the hamlet head. The *Gago* is assisted by *Madakis* who look after the hamlets comprising the village, it being possible to have several hamlets in one Gago area. In effect the Gago and Madaki structure was placed alongside the existing indigenous structures in Igalaland. As a result the offices of Gago and Madaki do not have an accompanying land entitlement as with the Attah and his traditional district heads and chiefs, but they are paid by government and are effectively civil servants although remuneration is minimal.

The biggest political unit in Igala, before the emergence of the centralised Igala kingdom, was the district in the modern sense. Boston (1968), confirmed that these traditional districts (called Ane or land) were less than twenty square miles in size and had a population of just a few thousand. Most of these traditional districts were regarded as large village areas and given Gagos. These Gago areas were grouped to form modern districts, and the present Local Government Areas within the state comprise a number of these districts. A consequence of this 'parallelisation' of governance structure is that the Madakis and the Gagos are closer to the people geographically, culturally and socio-economically and experiencing similar constraints. They therefore tend to have more influence than does the District Head or even the Attah, and when the Madaki or the Gago are involved in an action it is more likely to be acceptable to the people. This point will be returned to in Chap. 4 when the selection of participating households is considered. These traditional village administrations are supported by other indigenous institutions, especially in the form of self-help organisations (Seibel and Damachi 1982). The Adakpo (also known as Ayilo and Owe) is one such Igala self-help organisation. A degree of interdependence always existed within Igala society, with each person accepting some responsibility for the overall well being of the community. In the case of the adakpo, all farmers had the possibility of engaging in traditional rotation work groups in order to overcome seasonal work pressure in agricultural production. The number in the group varied. The person who engaged the working group on his/her farm had the duty to work for the same amount of time on all the farms of the group members. The adakpo was not a stable group and convened only as the need arose. In some places the female adakpo consisted of wives, relatives, neighbours and friends from the same area. In Ibaji and Odoru, largely due to Igbo influence, women do much of the farm work. Many own their own plots, having set aside sufficient capital to rent land. This is now becoming acceptable even in a patrilineal society. The same is true of other Igala areas bordering Idomaland. In the central part of Igala, women by tradition are protected from 'exploitation' in the form of heavy farm work. They are, however, allowed to own small farms and gardens where they grow vegetables and crops. Most concentrate on the harvesting and post-harvest activities such as food processing, cooking and the sale of surplus food in the markets. The adakpo system among women in Igala therefore varies from area to area depending on their traditions.

In the twentieth century the adakpo has adapted to changing values in Igala society. The introduction, during the colonial regime, of western culture as expressed through education, healthcare, cement and zinc houses and transport facilities had a marked effect on the aspirations of the people. No longer were young men only interested in going to the farm and young women content to stay at home all day. People now realised the power of money, and wanted to acquire it as quickly as possible. They became too impatient to wait for their turn in the adakpo as the rotating labour group might not come at the desired time. Therefore individuals with capital paid for labour when they required farm work to be done. For those without adequate capital, the adakpo is still important although members may carry out paid tasks for other farmers which in turn allow them to pay for inputs for their own farms. Presently the main tasks for which the adakpo is utilised are those that are relatively flexible in their timing during the growth of the crop, for example weeding.

Another form of traditional self help is the oja, which is another primary group operating at village level. It can be described as the traditional weekly meeting of a community group on an entirely voluntary basis. In rural areas these groups comprise the adult population, but in the urban centres youth who engage in labour tasks may also be in an oja. The Igala oja is very similar to the other Nigerian informal financial service institutions described by Eboh (1995) and Soyibo (1996). Many of these are based upon a rotating system. Eboh (1995) has suggested that in Nigeria the "majority of the rural population depend on informal finance". Each member of the oja makes a standard weekly financial contribution which is equal for all members. Members in turn receive an amount equal to the sum of all the savings for a week, less the cost of the entertainment and refreshments for the group for that day. Various other community responsibilities are discussed at oja meetings. Its traditional welfare system sees to at least the immediate needs of members who have suffered loss or bereavement. For this all members subscribe on a weekly basis and records are kept. As Igala were only partly drawn into the western system, traditional attitudes survived and the economies of women and men remained separate in the oja. Even today, women frequently form their own association as this allows them to dispose of their own income. These institutions remained untouched during the colonial period; it is only recently that these have been adapted to meet the changing circumstances without destroying their indigenous character.

The oja therefore plays a central role in regulating life in the village, it makes many decisions regarding the welfare of the community and responsibilities are taken seriously by the various groups allocated tasks by the leaders. It takes cognisance of both individual and group needs, and is facilitated by the presence of another community grouping known as the Age Grades Association (AGA). Above the age of ten, people of similar ages will be recognised as an AGA within the society. The primary function of these ongoing, aggregated age grades is to undertake community work which must be done all the year round e.g. maintaining hygiene in the hamlet. They are not paid for their services but are expected to do these tasks as a community service.

3.5 Igala Livelihoods; An Overview

The chief source of livelihood in Igalaland is agriculture, and Igalas are predominantly arable farmers with an average holding of approximately two hectares although this can vary dramatically. Cropping systems are complex, with marked differences created by soil type and areas of high and low population density. The systems are generally based on bush fallow, with fallow periods ranging from zero to ten years or more. The length of the growing season (approximately six months between April and September) allows for two distinct cropping periods, an early growing season (April to June) and a late season (July to September). A wide variety of crops, annual and perennial, are cultivated, and some keep livestock such as goats, sheep, chickens and ducks as a secondary activity. Livestock are often allowed to roam freely, except in areas of high population density where the larger animals are fenced in or tied up during the growing season to prevent crop damage. Apart from open cast coal mining in the Ankpa Local Government Area and the timber companies in Dekina and Idah, there are no other significant industries. There are some small scale businesses revolving around trades such as carpentry, block building and tailoring. Service industries, such as vehicle maintenance, transport, water supply and retailing, are also important, especially in the major towns. One person may engage in a number of activities, and farmers are often involved in a number of 'off farm' activities such as fishing, milling, maintenance and repair services, carpentry, tailoring, and trading. Conversely many people in employment e.g. civil servants may engage in farming in their spare time to supplement their earnings. Igala livelihoods are therefore highly diverse and provide exactly the sort of situation that SLA is meant to help understand and enhance.

The changes which took place in the Nigerian economy and noted earlier were, of course, reflected in Igalaland. While the value of the Naira was high in the 1970s the Igala farmers had to compete with cheap agricultural imports. High quality rice from Asia, for example, could be purchased in local Igala markets, and government owned stores sold imported food. During this period, many Igalas who would normally be engaged in farming found employment in the Anyigba Agricultural Development Project (AADP), road construction companies and at Ajaokuta steel works. The devaluation of the Naira in the mid 1980s brought many changes, and provided a much needed stimulus for involvement in agriculture. However, efforts from the government to enhance agriculture have been disappointing. There is little stability around the production of food and especially the prices which farmers and growers receive. The AADP was one of a wave of Agricultural Development Projects initiated across Nigeria in the 1970s as part of an African 'Green Revolution'. They were funded by the World Bank and were intended to spur the same sort of changes as had been seen in the Asian 'Green Revolution' of the 1960 and 1970s. The ADPs were in part designed in response to a fall in agricultural productivity noted earlier, and a concern to sustain domestic food supplies as labour had moved out of agriculture into more remunerative activities that were benefiting from the oil boom. Conversely, domestic recycling of oil income provided the opportunity for the Federal and State governments, with World Bank support in the form of loans, to develop the ADPs. The projects were of the 'integrated rural development' mode mentioned in Chap. 2 and provided agricultural inputs, extension services, health care, training, credit, rural roads, and village water supplies. The ADPs sought to increase food production and farm incomes primarily by the use of improved planting material and inputs such as fertilizer and pesticides, with farmers encouraged to move away from their traditional systems. But as it was recognised that farming households also required access to good transport links, health care, good quality water etc. then these needs were addressed in parallel. The first five ADPs in Nigeria were 'enclave' projects each restricted to a specific region within a state. Igalaland was selected as a site for an 'enclave' project in Benue State. This model of agricultural development demanded large amounts of capital and services and intensive management over a period of (typically) five years. With hindsight, not enough thought was given to the implications of the large increase in scale, or indeed to the less favourable production environment than existed in the smaller enclaves.

Nonetheless, for all of it's failing the AADP was a welcome project for the people of Igalaland. It began in 1977 and continued until 1982 when it was meant to go Benue state-wide as at that time Igalaland was part of Benue State. It was seen as recognition for the kingdom as a whole. However, as noted earlier this period coincided with the oil boom and as a result there was a shift away from agriculture to what was seen by many as easier and more gainful employment. For the five years of its existence the AADP provided such employment for many. Road construction offered many opportunities for labourers who were well paid. Security and labouring jobs were available; many of these gained from improved farm inputs such as fertilizers as they were close to the sources of inputs and information. Much observation and experimentation was possible at this level. The extension services sought to make inroads into places with farmers and all others with an interest in farming. Perhaps surprisingly, there was no attempt made to identify what it was farmers cared about or indeed wanted. The 'top down' nature of the project meant that the recommendations were designed based upon a blueprint of what crop production in Igalaland should be like and this was imposed upon the farmers. The result was that farmers took what they wanted out of the recommendations and adapted it to their own needs. Trials on maize and cowpea meant that improved varieties of these crops made their way into remote corners of the region. Large quantities of fertilizer were distributed to many villages, available for sale at highly subsidised rates. Compound fertiliser such as 15-15-15 were available at approximately £1.0 (equivalent) per 50 kg bag and Single Super Phosphate (SSP) for the equivalent of 60 pence. Other inputs such as pesticides were highly subsidised but apart from the control of spear grass (*Imperata cylindrical*), a particularly noxious weed in Igalaland, and insecticide sprays for cowpea (*Vigna unguiculata*) there was little interest. People living in bordering states availed of the opportunity for cheap fertilizer and improved seeds and greatly benefited. The road building component of the project was highly popular but also highly politicised. Villages such as Ekwuloku, one of the places selected for the SLA, benefited greatly from the road constructed near it through to Igboland. Similarly a number of borehole projects were established to bring clean water to communities, and these too were highly politicised, but unlike the roads the boreholes typically had a short lifespan and quickly became useless. The credit scheme was also not successful as repayments were very low and as a result it became unsustainable and discontinued. Even so many Igalas received training in a variety of skills and indirectly there was a boost to agriculture as some recognised the value of what was being done. But it was all happening far too fast and there was almost no participation by farmers in what the management of the project decided.

When the AADP went state-wide in the early 1980s Igalas lost many of the benefits it brought. Indeed the period between the early to mid 1980s coincided with the decline in Nigeria's fortunes and the subsequent introduction of SAP as well as the return of the military government. The ADPs were gradually re-absorbed back into the respective State ministry structures, and were substantially reduced in size and scope. The Igala gained from the AADP in its heyday but it would have been much more beneficial had it been initiated a decade later when so many Igalas lost their jobs and were free to be more disposed to farming. However, with the loss of the AADP and the cutting back of government services following the introduction of SAP, Igalaland found itself with almost no government support for agriculture or any other service. Indeed government services are now almost non-existent in the area and people increasingly dependent upon services provided by non-government agencies and the private sector. One of the most active of the non-government agencies is the Catholic Church (other Christian and Islamic groups are also active); running as it does a series of schools, hospitals, clinics, women's programmes and a rural development agency. This service provider is known as the Diocesan Development Services (DDS); its origins go back to the close of the Civil War and the birth of new ideas originating from the Second Vatican Council (1962–1965).

3.6 The Diocesan Development Services in Igalaland

DDS, the lead agency in instigating the SLA set out in earlier chapters, has its roots within the Catholic Church in Nigeria, and more particularly the greater emphasis of the Church on its social mission following the Second Vatican Council, The Second Vatican Council (Vatican II; 1962–1965) emphasised the Catholic Church's relevance to the situations in which it found itself, and it is not surprising that agriculture found a place in the new Church ministries that emerged in the wake of this council. With Vatican II, the Catholic Church could no longer be indifferent to the overall situation in which it found itself as the opening sentence of Gaudium and Spes (Vatican II 1963–1965) states clearly:

> The joys and the hopes, the grief and the anxieties of the people of this age, especially those who are poor and in any way afflicted, those too are the joys and hopes, the grief and the anxieties of the followers of Christ.

DDS was established in 1971, in the Diocese of Idah, some six years after the close of the Council. This diocese is co extensive with Igalaland and was formed during the Civil War in 1969 first as a Prefecture Apostolic, and later in 1978 it was raised to the status of a diocese. Idah is the capital city of Igalaland, and home to the Attah of Igalaland. DDS was initiated only two years after the diocese was created. Initially there was a piecemeal understanding of the problems and the general situation in the diocese; the situation was worse in the war affected borders as Igala shared boundaries with Igboland. In 1971 it was possible to single out a new middle class; a direct benefit of education programmes which all churches did much to promote. This social mobility was possible for those who could benefit from educational opportunities.

To effectively engage in the concern expressed, the first prefect Apostolic of Idah Prefecture, Right Reverend Leopold Grimard, in 1970, invited a full time social scientist, one of the authors of this book, to help examine the situation in Igalaland with a view to responding to the calls of the Vatican Council. The initial vision was to spearhead the process that would lead to an escape from poverty and the dependency which it creates and fosters. The underlying principle was to help each individual to achieve full personal potential by creating a situation of greater opportunity to the greater development of the community as a whole. From the beginning, self-reliance and development from the foundations upwards was the underlying philosophy and which was always encouraged. Because of the many contacts already existing through parishes, schools and clinics, this 'foundations up' development would be possible. The essential movement had to flow from these primal contacts and had to be grafted onto indigenous institutions, such as the *oja* in whatever way these could be adapted to meet pertinent local needs. However, progress in such a framework was bound to be slow, as on the one hand genuine needs had to be identified by the people themselves, and on the other hand the process had to sensitise itself to the socio-cultural integument of Igalaland to detect that precise place in the natural order of things, where a solution might be coaxed into existence, almost like a skin graft. Dialogue and listening were the *modus operandi*. The pitfalls were many and progress could only be made slowly. Also, an occupational hazard of dialogue as the idiom of successful grafting is that those who are better off and more articulate can present their needs with more clarity and urgency and, as a result, make off with the lion's share of what is available. Development then can easily become an agent which increases the gap between the rich and the poor, and what often happens so blatantly at national and international levels can become a model of the same thing at local level. Because of such concerns and in the light of these ambiguities, it is difficult to establish a neat and clear criterion with which people's responses are taken into account; this further complicates the issue. Not only do basic needs differ, but responses differ infinitely, especially in situations where outside aid is offered as part of the solution to the problem. Development intervention must, in many ways, be left flexible. This very flexibility makes the implementation of agreed projects more difficult and slows down the rate of progress. Once ideas circulate, each individual interprets these according to his/her basic needs, expectations, experience, vision and background. Fear can colour the whole situation: many fear change and feel more comfortable with the *status quo*. A disruption could change conditions drastically, especially when leadership and unjust social conditions are challenged, which frequently they must be.

With this in mind, the newcomer to Idah Diocese charged with this task realised that an 'in-depth' understanding of the Igala people, their socio-cultural structures and organisations, would be necessary before attempting any development work. Such a daunting task indicated how carefully one had to tread within such sacred space. One needed to ponder and be aware of the respect needed. Such approaches and attitudes and of course values form the very foundation and cornerstone for sustainable development and indeed the basis for an SLA. The first objective was to contact all organisations (with special attention to women), in the diocese to ascertain their views, aspirations and suggestions regarding individual and community needs. This was facilitated by the extensive diocesan network which had contact with all segments of the Igala society, irrespective of religion, social status or political opinion. These contacts provided a good insight as to what developments might be possible. But if people were to be the subject and object of their own progress, short cuts were not possible. The emphasis was shifting to a careful examination of the overall structures of Igala society. This was explored with local historians as well as church committees and the many primary groups already mentioned. Generally this 'Cinderella like' kingdom was depressed, due mainly to the factors already discussed but especially the Civil War, and there appeared to be little money in circulation. The road network was in poor shape, and other than a limited number of taxis, public transport was nil. Women and children walked three to four miles to the stream each day as water availability was a serious problem, with the procession usually commencing long before dawn after which children often had to walk many miles to school. Many were too tired to be attentive if they went at all.

At the invitation of some villagers, the development office (now called DDS) extended its dialogue to villages of Ofakaga and Ejule. This time there was one need agreed on by all members of the society; namely water. The women in particular, but now with the approval and consent of the male population, agreed to raise funds for a water project. As this could only be done through the sale of their agricultural surpluses, dialogue on the topic inevitably led to discussions on profits from agriculture. While water continued to be an obvious priority, the farmers soon became as interested in agriculture as in the water project itself. The fundraising required a piece of land where crops would be grown and the proceeds used for the water project. To get the maximum benefit, the villagers invited the staff from what was then the Kwara State Ministry of Agriculture (MOA) to help. The Ministry had already been alerted to this maize growing venture, and treated the maize plot as a type of pilot scheme. Governmental agricultural policy at that time was promoting the use of fertilisers to increase crop yields, this being part of their modernisation program already in vogue in the country and which would lead to the birth of the ADPs in the mid to late 1970s. The plot was arranged so that the effects of fertilisers could be tested on maize in a manner that would show the differences between fertilised and non-fertilised areas. A Muslim farmer volunteered the land; others provided the labour and seed, DDS the fertiliser, while staff of the MOA accepted responsibility for the technical expertise. The maize was sown in May 1972 and harvested in September of that year. Rainfall was favourable and normally a good yield could be expected. However, the yield from the fertilised plot amazed everyone, even the MOA staff who had similar trials in other locations. Previously, some fertiliser had been distributed by the MOA, but was used by women to whitewash their houses! Local farmers decided that, henceforth, it would be more beneficial to use the fertilisers on crops, a decision regretted by the women.

At first, the demonstration was viewed by the participating farmers, even by those most closely involved, as being somewhat amusing, until the high yield from the fertilised plot jerked them into taking the matter seriously and realising the worthiness of engaging in further research. Non-participating farmers were still sceptical and somewhat suspicious of agricultural innovations, and considered that future experimentation should be with other crops of their own choice. Excitement over the successful demonstration heightened the importance of agriculture in the locality, resulting in the same farmers highlighting other constraints to increasing their agricultural production. Foremost amongst these were:

1. An acute shortage of money for agricultural inputs
2. The high labour requirements in general especially for land clearance and initial cultivation
3. The difficulties of weed control
4. Low soil fertility
5. Poor quality seeds and planting materials.

From a tight-lipped conservative reluctance to ever mentioning their farming problems the floodgates suddenly opened, with the feeling that they could never get enough outside help and co-operation, an event warmly welcomed by the MOA staff. This process gave rise to hope for future improvements, but these developments progressed at a slow, though encouragingly steady rate. The importance of dialogue at many different levels was apparent and nothing could be taken for granted.

The next stage was to encourage these farmers to consider establishing a structure enabling them to continue the process in a more organised manner and to decide their priorities in view of the many problems now evident. This resulted in the adaptation of traditional structures to meet farmers' needs. However, it is interesting to note how agriculture, as a means to an end, eventually took centre stage and not the initial concern which was water. Indeed, for reasons beyond the control of the villagers and DDS, it took many years before the initial dream of an improved water supply would be realised. This dream was not allowed to fade: water supplies so vital to sustainable livelihoods are now possible in a highly sustainable manner. Local methods have been upgraded; square or oblong tanks have been replaced by oval shaped structures that have no joining.

At subsequent meetings in Ofakaga and Ejule the scarcity of money for agricultural requirements was again highlighted, and appeared to be at the core of every anxiety. Given enough cash, more labour could be hired and good quality planting material and other inputs purchased. The monetary contributions inherent in the oja seemed worthy of consideration as a means of accumulating capital for agricultural purposes, and farmers were asked what they thought about the idea. Their immediate reaction was again apprehensive. But now with the help of dialogue it was possible to progress from reaction to a more satisfactory response especially in relation to a topic introduced by the group for a solution. In this instance they consulted among themselves, and their reaction was to request DDS to help convert their contribution system into a form of agricultural finance that could be used by them at the commencement of the rainy season when money was scarcest but most needed. It was realised that there were problems of adapting it in its original state. Foremost amongst these was the fact that all members would require their money for planting at the same time, and could not wait for their turn on a rotational basis. Another common problem with the weekly contribution system is that records were not always kept, and it was also possible for the money to be stolen. It was necessary to guard against such an eventuality.

As already outlined the oja system had many advantages. One of the most important of these was its organisational structures at village level. The initial groups in Ejule consisted of four hundred members, some of whom were women. This group was therefore different to the regular village oja in that it was larger and consisted of both women and men. The idea of dividing into smaller groups arose in the course of the discussions between farmers and between farmers and one of the authors. A number of meetings ensued before it was eventually concluded that the members of the larger group divide into smaller ones. The members of the larger group decided that each small group to be known as Farmer Councils (FCs) or sometimes as 'home councils', comprised 10–25 members using the well known hierarchical organisational structure of the village oja. Chairmen and secretaries were important functionaries in the FCs. The choice of chairman was usually associated with age, a fact that is not surprising since seniority is respected in the patrilineal society. In the case of the secretary it depended on literacy. Each FC met on a weekly basis and all chairmen and secretaries attended a monthly meeting that was chaired by the zonal chairman. Chairmen from the FCs formed a zonal group that met once monthly to assess overall progress. The presence of gagos and madakis was an essential ingredient in the evolution of the FCs given their connection with local people.

With the FC and zonal structures in place, members requested that further adaptations were required, especially in relation to their financial requirements. The first step towards solving the problem of cash shortages for farming was achieved by adapting the oja contribution system to a savings scheme. The scheme was based on each farmer making an agreed weekly contribution during the year, and at the next planting season the FC withdrew their savings. However, one of its major drawbacks at the time was the absence of records of money saved, so accountability needed to be built into the system. As this contribution system resonated with that of the Credit Union (CU) (one of the authors had extensive training with the CU movement) ideal of getting control of one's own destiny by saving regularly, the advisability of adopting its accounting methods seemed to make good sense in this particular situation. With the assistance of Igalas trained in the CU accountancy systems, forms were designed which could be used by the groups to record their weekly savings. Strict fines were imposed for failure to attend, as it was believed representation was no compensation for participation and many new ideas were discussed at the FC level. Problems were discussed and decisions made. Continuity was an essential ingredient to grasp what was taking place.

DDS in turn had to set up a system that had full records of all savings mentioned above. But while the introduction of a savings scheme was an improvement, it was clear that the sum saved needed to be augmented. The FC members were happy to put money aside for farming, but unfortunately the amounts saved were meagre when compared with what was required at the commencement of the rainy season. Clearly, the provision of appropriate credit to farmers was a basic need required by the FC members before they could even consider any programme of agricultural development. As a solution, the idea of a loan scheme of the type given by the CU surfaced. The Igala staff, who had become familiar with these procedures, believed the DDS could also adapt them to the needs of the FC. Until 1973, apart from the salary of the seconded staff, the programme needed no external input. However, a decision to augment savings by a loan would require a departure from self-reliance. DDS by this time felt that an advisory committee would be required for decisions such as this, and DDS was in constant communication with the church network described earlier in the chapter which greatly facilitated dialogue and feedback. A committee comprised people from these contacts along with others who might have a different opinion based on different backgrounds. Selecting a committee did not prove to be a difficult task as many educated Igalas had taken an interest in what had already happened with the saving scheme, with some teachers adopting the practice at their schools. The committee consisted of some Canadian missionary priests who had worked in the locality for many years, most with experience in the development of the diocesan education programmes. Other committee members could draw on years of experience, either in the public or private sector, most of which was applicable to the situation. They visited the FC meetings and discussed with members. After much debate it was decided that outside assistance would be sought. At that time, the Catholic Secretariat of Nigeria (CSN) had a fund available for such initiatives, so the actual procurement of the finance required was not a major hurdle. Far more important was making the decision to undertake a loan scheme, as everybody was familiar with its hazards. It was often perceived as a 'dash' (Nigerian term for a free gift), which would not require repayment. However, as a saving scheme was effectively in place, loans could be organised using it as collateral. It was the practice of the CU at that time to give a loan to members equal to double the amount saved, and charged a small interest on loans. The advisory committee felt that these two practices were sound and related to the specific needs and ideals that were important to the FC members and DDS. However, getting an almost completely illiterate group to understand the complexities of interest calculations was a mammoth task, and it was eventually decided by the committee that the rate of interest would be one kobo per month on every naira borrowed (equivalent to an annual interest rate of 12 % as one naira = 100 kobos). Receiving double the amount saved would at least be an improvement on what was available to members at planting time. The committee was also adamant that the loan was to be spent only on agricultural inputs. This had to be carefully monitored.

The decision to take loans resulted in the formation of strict rules by the FC members. The zonal and FC chairmen and secretaries looked seriously at group membership, especially as it was made very clear to them by DDS that they would be the people responsible for seeing that repayments were made within a year. From then on their positions would be onerous, rather than prestigious. They had to be satisfied that their members were credit worthy, and some members opted out as soon as they realised there would be no free handouts. Quite a bit of reorganisation took place at FC level to guarantee the cohesion essential for smooth operation. This was at their own instigation as they were anxious about their reputations and future help. It was clear that DDS had only limited funding, and required some guarantee that its capital would be returned. Loan contract forms were designed giving legal status to the agreement being entered upon by the FC and DDS. This contract was signed by the FC chairman on behalf of their members and by the zonal chairman as a guarantee that the FC chairmen would repay on time. It was also signed by a DDS representative. However, distribution of the money between members was the FC's responsibility. One of the first rules made by FC members was that the repayment of the loans would be made through the FC chairmen. If the loan was not fully repaid before the next planting season, a further loan would not be forthcoming. There was a high degree of dialogue between FC members and DDS in the creation of the operational framework within which the scheme operated. The DDS had its responsibility to maintain a fund with the interest required for administration. Yet it needed the flexibility that took into account farmers' constraints.

Repayment of loan and interest as a one-off payment could mean that a farmer had little capital left to reinvest in farming. To circumvent this eventuality, members were encouraged to repay the loan in instalments. For example, repayment on a monthly basis reduces the hardship of repaying capital and interest all together, and repayments could begin as early as possible. It was, however, seen that keeping surpluses for as long as possible would mean extra profit if storage facilities were adequate. Such decisions rested with the members.

The Canadian priests based in the diocese took a keen interest in the savings and credit programme as it developed and thanks to their presence at zonal meetings the accounting system was almost impeccable. They often took the money for safe keeping before handing it over to the DDS office. They maintained the accounts with such precision and orderliness that high standards were set, providing a good example for the DDS office. Since some were also members of the advisory committee and active in the parishes, they helped preserve the accuracy of much of the discussion material that needed to be thrashed out by participating members. They knew the language and idiom and one wonders what the programme would have been like without them. They had always welcomed an opportunity to engage in agriculture and felt privileged to help in a manner they believed most relevant to the lives of the Igala where agriculture was common regardless of creed. To make the scheme self-reliant a small administrative cost was made by each FC to help with transport costs, expenses of DDS staff and some administration. Committee members, from their experience of the school system, suggested slightly higher fees as giving more credibility to the institution in question as it showed this type of operation cost money to run. A few members did object to this local contribution or 'community contribution' as they themselves referred to it. So vigorous was their outcry that one could be of the opinion that the loans were for the benefit of DDS and not the members themselves.

Monitoring of farming activities was central to success. Intensive supervision by DDS resulted in higher levels of FC participation within the areas in which it worked. The experience showed DDS that farmers liked to be checked with regard to loan usage, and were able to clearly identify the exact items on which expenses were incurred. Indeed, they viewed the role of DDS as one of encouragement rather than interference, and appreciated the fact that both borrowers and lenders needed to have a good understanding of loan usage. Given this interaction, DDS could equally learn about the drawbacks as well as the benefits of credit in this context. Too small a loan was of no benefit in terms of increased productivity, yet too high a loan could make farmers dependent for a long time. Close contact with loan beneficiaries was crucial for DDS in its reflection on loan impact, and facilitated flexibility and scope for experimentation. Unmonitored credit, even if based on savings as collateral, can very well have the effect of destroying self-reliance. DDS was at times concerned about its level of contact and supervision and wondered if this was perceived as paternalistic/maternalistic. The Igala reaction was a feeling of status which encouraged participation and a belief that the DDS had a genuine interest in their livelihood. DDS, however, took a risk here because of the possible perceived negative reaction. In hind sight it was a risk that paid off as it allowed quality dialogue to evolve. It was seen as respect and an acknowledgement that they were the *raison detre* for the work. But not all groups worked with DDS in the same way. Generally DDS responded to requests for assistance and projects then undertaken in dialogue with that community or group in the spirit of self-help. Every parish initially was made aware of interventions in various places and dialogue took place at different levels within parishes as to what help was possible according to needs but especially in relation to agriculture. DDS was ever conscious of equitable distribution of opportunity. But people and groups respond differently and the one size did not fit all needs. For a new comer to DDS, it might look as if Bassa Komo was not assisted to go the sustainable development—SLA route. However, considerable work was done in Bassa Komo initially (1970/84) with infrastructure such as bridges. The savings and credit scheme was also introduced and people trained for this work; but despite dialogue and encouragement there was not an understanding of self-help development in an area rich in agricultural land and almost total dependence on farming. But Bassa Komo received considerable help with nursery schools, water projects and a primary health care clinic (1980 and 1990s) but was not ready for anything approaching an SLA. Bassa Nge was further away but it was well recognised that apart from not seeking help it had well developed linkages throughout Nigeria as Bassa Nges were among the first to be educated. The Anglican Church has always been active and over the years good relations developed especially in relation to health and nutrition services. It is a question of reciprocity and exchange of ideas which is enriching for all parties. In the post—war era DDS helped with rehabilitation services along the Igbo borders. As Missionaries from here were interested in agriculture good progress was made initially. But as Missionaries decreased in number, and roads deteriorated and due to language problems many groups developed from a certain point to be more self-reliant. Some (including Ekwuloku) came to DDS with different problems to be resolved and where they knew they would be helped if possible. DDS never encouraged dependency. Finding suitable staff was also a problem as demand for services in active areas increased. Many came to the DDS office for assistance and an outreach was not always required.

Another observation is that the rules made by the FC members were strict as they wished to ensure that no member shirked their responsibilities to the FC. Within the FCs there was little social differentiation, as a high level of economic homogeneity existed between members and between FCs. A stranger might be a potential saboteur and hence not allowed membership, which is so highly based on trust. Because the DDS was an organisation committed to development it was motivated by the desire to generate greater equality between the different households and families. At first DDS did occasionally insist that certain applications that members deemed unsuitable be honoured, but this was disastrous. The result was that the remaining FC members withdrew, leading to the loss of valuable initiative and capable leadership. With time and experience DDS eventually found that it was possible to design other plans with those not deemed worthy of FC membership. This took time and a good deal of research and intensive dialogue. It was clear that any one initiative is not all embracing and development intervention has to be keenly aware of the shortcomings innate in the most conscientiously planned scheme. For example, women were always active in the FCs though numerically less than the male membership. One of the frustrating events that occurred in a number of places was that when the women excelled in any way, men could decide it was time to close down the zone, and all this without any dialogue with DDS or neighbouring zones. When eventually the situation became clear, antennae were sharpened to greater awareness should a repeat performance occur. Occur it did and it became the beginning of new insights as to the differences between the female and male roles in the programme. Understanding the specific and often complex networks within households is essential to sustainable livelihoods.

As staff numbers increased it became clear that sharing a vision is more complex than simply supplying the skills for performing a task. There are two separate issues at stake; a vision and technical skills. A vision can only be internalised and bear fruit when it is eventually owned. However, technicalities such as record keeping can be taught. Agricultural development needs innovation, care and planning. Without the vision the concept of self-help can become a mere word or an isolated punctum if the philosophy is not kept in sight. Fortunately, the philosophy was gradually grasped, but not without its many critics. Technical skills can never be underestimated, and it is the combination of the vision and skills that brings good ideas to fruition. Without such a merger, good ideas can be reduced to bad experiences.

As time passed, the conditions for becoming FC members and the organisation of the FC programme became clearer. It did not, after all, promote or rely on imported knowledge. The operational framework was based on traditional institutions, with FC and zonal chairpersons selected along traditional lines. The founding members were extraordinarily generous with their time and explanations in assisting their interested neighbours. They were keen that a movement which they had begun so successfully, according to them, would retain its respectability and utility. At no point did DDS advertise the FC programme. It had become embedded in the affairs and the very fabric of the society and essentially self-generating.

Another important factor that needs to be mentioned here was the relationship between DDS and partners-donors. Links with the Ministry of Agriculture have already been mentioned, but partnerships went far beyond this. Over the 40 years or so of its existence DDS has sought support from a number of international development donors, and as perhaps would be expected most of them existed within the broad family of the Catholic Church. Notable relationships were with MISEREOR (Germany), CAFOD (England Wales), Trocaire (Ireland), CARITAS (Holland), Development and Peace (Canada), Bridderlech Delen (Luxembourg) and Missio (Germany) and the Little Way Association (England). But there were others that were not faith-based in orientation such as Oxfam, Self-Help Development International (Ireland), RASKOB (USA), Gorta (Ireland), Electric Aid (Ireland) and the Combined Civil Service Third World Fund (Ireland). Some government aid was also provided, most notably from Irish Aid, CIDA (Canada) and DFID (UK); various embassies also helped. All these agencies had their own especial vision and mission as well as processes and requirements that had to be met and managed by DDS, but they also provided invaluable support and insights for a range of projects including micro-finance. It was, and still is, a rich landscape of partnerships that changed over time as some donors left the scene while others came to the fore. Some of these partnerships involved funding for research rather than 'development' work; the lines between these two were often blurred as DDS engaged in 'action research' whereby communities sought to establish what was needed in terms of an intervention.

Finally, it must have occurred to the reader that during the late 1970s and early 1980s the AADP had a strong presence in Igalaland then how did this affect DDS? These two organisations were about as far apart as they could be in terms of how they worked but following an extended dialogue between AADP and DDS it was clear they did share the same basic goal of wanting to improve the livelihoods of Igala people. The AADP manager stated clearly that DDS would continue when AADP had moved on so in the interest of adding value to what DDS was doing every effort to cooperate for the maximum use of resources would be made. For the five years of the AADP FC members became contact farmers for the AADP extension service which was headed by one of the authors. DDS staff temporarily co-opted to the AADP. There was monthly up-skilling which was helpful for staff and farmers. When the AADP went state-wide in the early 1980s DDS carried on with what it had been doing before the AADP arrived. However, there were complications. Notable was that farmers had become used to low-cost and often free handouts such as fertiliser and pesticides provided by the AADP and were understandably chagrined when expected to pay for them. Indeed in many cases these services were no longer available, and DDS as the only functional agency left in the area took the brunt of the complaints. At that point in time (1978–1982) Nigeria was enjoying the oil economy and many partners withdrew assistance from Nigeria. One partner showed one of the authors the massive reduction there had been in requests for projects from Nigeria. It was unfortunate for Igala that it was when the AADP was no more that SAP reared its ugly head. This situation was further compounded by some donors who believed that the Igala had bene-fited enough from the AADP and there were other more needy groups than were in Igala. Add to this mix the complication of a military takeover in late 1983 and the sanctions imposed by foreign governments who did not agree with what the Nigerian government was inflicting on its people. It was a turbulent time. When help was needed most DDS competed like many other groups in the same situation for scarce funds. It was time to diversify and find more channels of help.

3.7 New Pastures

DDS had worked with tens of thousands of households over more than four decades since its inception in the early 1970s, reaching a minimum of half a mil-lion people over four decades and had a proven track record which earned it the trust and respect of many partners and especially local Igala communities. But unless there is adaptation to on-going change, organisations will outlive their usefulness. DDS insisted on internal and external evaluation so as to measure its effectiveness and its future orientation in the evolving context of Igala and Nigeria. DDS has been involved in a wide range of programmes over the years mainly because as a result of learning from evaluations and updating of skills in order to maintain the vision that guides it. DDS, on reflecting on over three decades of credit provision, decided it was time to move into a more business-orientated provision of financial services and though initially an internally-led change it was welcomed by many participants. The drivers for this transition were multi-fold. To some extent it mirrored what formal lending institutions had been doing for some time and thus was becoming more familiar. Also, DDS had some defaulting on its loans, which when investigated were typically failures by borrowers to use the money for the purpose for which it had always been given—an investment, Instead it may have been used to pay off another loan, debt or used for another pressing problem such as illness or death. The difficulties faced by people during the 1980 and 1990s following SAP and the political and economic turbulence have been set out in previous pages. DDS saw the need for a new orientation regarding its credit conditions to ensure that borrowers had the best possibilities to improve their livelihood as well as repay loans. Thirdly there were changes amongst the inter-national donors who supported DDS. These relationships were long-term despite certain ups and downs. There were times when given the political situation in the country donors were reluctant to be seen supporting agencies in the country; even if they were Catholic Church based. The decline in the importance of the agri-cultural sector in Nigeria also corresponded with a general retreat by donors for funding agricultural projects. DDS had to adapt to these changes and reduce any need for outside funding for its micro-credit scheme. But this transition was by no means sudden and comprised a series of changes over time. There were some protests and dealing with these took much time and patience. But it has to be remembered that donors also go through their transitions with change of management and governance especially if there is co -funding and have to justify the funding of programmes in the same place for many years even if the actual programmes changed. This is healthy. A possible example of such reflections is the more recent demand by partners for a business plan now required for income-generating projects. In an era where sustainability is vital, business plans are becoming essential ingredients in demonstrating that a programme has a future after initial funding. An illustration of this in DDS was a decision in 2002 to engage in a proposal to DFID for funds to develop new methods of clean seed yam production with a business plan as an integral component. Despite its decline amongst national policy makers and international development donors the importance of agriculture in Igalaland has already been stressed. It was about to have a new incarnation responding to a need for healthy seed yam in an entrepreneurial environment.

Within this maelstrom of change DDS became involved in a series of research projects funded by DFID and focussed on white yam (*Dioscorea rotundata*). Yam is a major crop in Igalaland, and one of the major factors limiting its production and revenue is the availability of healthy planting material. Yam is a vegetatively propagated crop, like the Irish potato (*Solanum tuberosum*); the planting material comprises small whole tubers or pieces (setts) of a larger tuber. While this has advantages such as the consistency of valued characteristics in subsequent generations, it does mean that pests and diseases will also carry over from one generation to the next and can even build-up within the tubers. These reduce yield and quality of the material which in turn reduces its market price. The DFID projects implemented between 2001 and 2005 aimed to identify low cost alternatives that would help farmers control these problems, and combined a series of research-station based experimentation with farmers surveys and on-farm (farmer managed) trials. DDS were involved in the latter two activities—the surveys and on-farm trials—and the decision to include them is easy to appreciate given their extensive presence in a major yam growing area.

Given difficulties in securing funding focused on agricultural issues of importance in Igalaland, the DFID projects offered a unique and excellent opportunity. DDS saw the DFID project funding as allowing for progress upon a number of inter-related fronts. Firstly there was the obvious potential for promotion of clean seed yam production as an important system in itself, largely because it could serve to increase the quality of seed yams for many farmers and ultimately enhance livelihoods since yam is such an important commercial crop as well as prized for consumption. It was hugely advantageous therefore that the DFID project aimed at addressing important issues for yam growers in Igalaland. Secondly DDS saw an opportunity to 'lever' other benefits from the project. DDS felt it could use the project to expand to areas in which it hitherto had had little engagement, especially the border areas between Igalaland and Igboland from where people at different times came to DDS for advice on different matters e.g.

agro-forestry but since this contact was rather sporadic this was a chance to change the situation. Farmers trained on the DDS farm sold their seed yams in these areas but this intervention had more potential for DDS to be catalysts for change. The people of those areas are mixtures of Igala and Igbo and the culture is quite differ-ent to other areas where DDS had traditionally worked. But yam was known to be an important crop there even if it was also known that illegal yam cultivation took place within nature reserves because of the better quality soils. Thus the DFID seed yam project looked like an excellent opportunity to engage with those people. Also, provision of better quality seed yam was known to be beneficial to the wider population of yam growers, but for this to be sustainable the seed yam producers also needed to benefit. This made it advisable to explore the livelihoods of farmers engaged in seed yam production and how this could impact upon them. In part this involved an economic assessment of the on-farm trials, where the treatments could be compared in terms of economic costs and revenue, but it was felt that clean seed yam needed to be seen within the wider range of livelihood options and con-straints that farming households faced. Linked to this, of course, is the potential for microcredit to help such seed yam producers and this fitted neatly into the ongo-ing reflections within DDS towards a more business-plan approach to supporting enterprises. After all, the clean seed on-farm trials could be considered as an enter-prise that many farmers might wish to engage with if they had financial support while at the same time affording DDS a facility to learn much more about the live-lihood context of this enterprise and use that to develop the financial support that might be needed. Indeed the broad lessons that DDS expected to learn could be applicable to a wide range of enterprises and not just clean seed yam. DDS also saw potential benefits of using the DFID projects to lever more funding from other donors, especially for proposed changes with its microfinance scheme; something which DFID was keen for DDS to do. As the DFID projects were more technical and crop-specific in nature, although still involving farmers to an extent, DDS felt that providing a broader and more human face to the work would resonate bet-ter with other donors, and an SLA would be a useful approach to adopt given the widespread (at that time) use of the term by these donors.

But while the DFID funded seed yam project provided a number of opportunities, DDS was well aware that implementing an SLA in those places where the DFID project was working would not be an easy task, especially with the limited resources at its disposal even with the additional funding available from DFID. This was carefully thought through and the decision was made to explore the livelihood of those households involved in the on-farm trials rather than attempt anything on a larger scale. The advantages of such a focus would be as follows:

1. The logistics could be more readily controlled and costs kept down as a result, Part of this included a decision not to engage expert help with the SLAs as that could be expensive. Instead DDS sought the advice of those already involved in the DFID seed yam project.
2. A much more in-depth SLA could be conducted rather than something that might engage a lot more people but could be quite superficial. The value of the in-depth study was that it would involve a longer-term relationship with the

selected households and provide an opportunity for trust to develop, The in-depth engagement with a relatively small number of households could help lead to their involvement as catalysts for engagement within the wider community.

3. As will be explained later, there were also concerns about the quality of data with larger-scale studies. DDS had much experience with such surveys implemented by outside agencies, including evaluators funded by various donors to explore the impact that DDS was having in the area, and was well aware that larger-scale studies tend to provide more superficial pictures and there is a significant danger of acquiring poorer quality information.

4. The in-depth engagement would readily allow for a more qualitative-based data collection via regular discussions and observation of change over time. DDS was aware from its own experience that much value could be lost with larger-scale and more structured processes of data collection that tend to reduce complexity to relatively few numbers.

In effect these factors resulted in a case study approach to SLA which would focus on only a small number of households. They inevitably resulted in a set of trade-offs designed to best match what DDS felt it could do and what could be gained. It is important to note that these were not decisions made in an abstract sense of providing the best quality data for scientific publication or to provide the basis for more in-depth research that would lead to such publications. While this book presents the outcomes of the SLA and what surrounded it, the SLA was not implemented to produce the book as an output. Neither was the SLA intended to provide an evidence-base for a range of options that could be taken to state or federal government or indeed other agencies for them to 'do something' to help. The context was much more grounded within the community which DDS sought to serve and will continue to do so as long as the Catholic Church is present. On the positive side of the balance sheet it could reasonably be assumed that trust was a major advantage in any SLA and the involvement of DDS seen as beneficial by those communities selected for engagement. Igala recognises DDS as part of the fabric of its society; though not linked to government directly, government and DDS cooperated for their mutual advantage and to the benefit of the communities. Secondly, the grassroots mode in which DDS has functioned since the 1970s means that its Igala staff has much experience of working with Igala households. They share the same culture with its extensive knowledge of language, idiom and livelihoods. Such advantages give opportunities that could be considered ideal in conducting an SLA. If it did not succeed here it would be difficult to imagine where it would.

3.8 Choice of Villages for the SLA

The farmer-managed trial component of the DFID seed yam project required the selection of eight trial sites in total within Igalaland. Given the resources available it was decided to concentrate the trials at two villages; Ekwuloku and Edeke

(Fig. 3.1). There were a number of reasons for this selection, and they involved something of interplay between choice of location for the yam trials along with the other and wider concerns noted above. Edeke certainly fitted the bill as an excellent site for farmer-managed yam trials given its rich riverine soil which is ideal for yam production. Indeed households there have a strong reliance on yam (*Dioscorea rotundata*) for income as well as subsistence. Edeke is relatively close to Idah, the capital of Igalaland, and the headquarters of DDS, factors that would help simplify the logistics of the trials. Edeke was also representative of the many villages in the Ibaji District of Igalaland, south of Idah, whose involvement with the DDS microfinance scheme dated back to the 1970s. Hence DDS knew that a high degree of trust existed and this would help facilitate the SLA. DDS also wished to include a village from the plateau region of Igalaland, inland from the two main rivers (Niger and Benue) but one which had less involvement with DDS but representative of an area for potential growth. The village also needed to have a strong interest in yam, although in the plateau region the cropping patterns tend to be more diverse. Ekwuloku fitted this requirement, and like Edeke was relatively easy to access from Idah. Ekwuloku is close to the Igbo border and therefore hosts farmers, traders and others in search of land and opportunities scarce in their own state (Enugu State). As a result the population of the village is a mix of Igala and Igbo, with Igbos in the majority. Almost all speak Igbo but there is also a dialect of Igala. English is their second or even third language. The farmers in Ekwuloku do grow yam and the village has a significant market for the crop. One other factor that resulted in Ekwuloku being selected was the discovery that some farmers there were using a local forest reserve for growing seed yams. This is an illegal activity in Nigeria but did help sway DDS into selecting the village so it could help wean farmers from cultivation within the forest reserve. The following paragraphs provide some more detailed background for the two villages.

Edeke lies a few kilometres south of Idah along the fertile flood plains of the Niger River, far too close for it to be discernible on the map in Fig. 3.1. It was accessible by road only during the dry season but river transport comes naturally to Edeke inhabitants; however there is now an all season road from Idah to Edeke. It has a population of approximately 40,000. The six clans that own the land can trace their ancestry back 400 years to the time when they were indigenes of Idah. The area now consists of nineteen *Madaki* (chieftaincy) areas. There are many migrants now living in the village, a good number of whom hail from the wider Igala area. Non-Igalas include Hausa, Igbo, Fulani, Yoruba, Nupe, Urobo, Isoko, Igbirra, Kakanda and many others. Migrants found their way to Edeke mainly because of its potential for farming, fishing, hunting and livestock. While not so important now, cattle were once central to the Edeke economy and are still remembered by local inhabitants; one of its islands is called after the cattle and recently a primary school built there was given that name Alla Okwuno (The Cattle Island). Edeke has a natural divide—uplands and flood plain—that mirrors the divide of Igalaland as a whole. The flood plain is completely flooded during the rainy season while the upland area enjoys good soil moisture though it is never completely flooded. The main crops grown are yams, sweet potato, maize, pepper and small areas of rice. Agriculture is practised in the uplands along with hunting to supplement income and protein intake; fishing takes pride of place in the lowlands. The annual flooding of land replenishes the soils and yams can be grown continuously (mono-cropped) in the same site for as long as six years and given their relatively high market value the crop is understandably the most popular. There is an established rotation system in the uplands—cereals to root crops to green legumes to vegetables and on to fallow—and there is intense cultivation with more land planted in the early part of the growing season before flood waters rise. Cultivation especially of yams, rice and vegetables begins in November which is much earlier than for the plateau area of Igalaland where the planting season begins with the onset of rains in March/April. Women are engaged in rice production and the money earned from it used by them as credit to male yam growers at an interest rate in the region of 100 %. In theory this might appear good for the women but in practice can be problematic as the male borrowers have difficulty in repaying and this is now a persistent problem in Edeke. A limited number of trees are grown in the uplands, which the lowlands cannot support due to seasonal flooding. Oil palm is the most popular though for the most part these are self-seeded rather than planted. Fruit trees especially the improved varieties are now to be found especially oranges, guava and mango. Calabash trees are also popular due to their role in the local fishing industry. Bamboo can be found everywhere and is considered vital to the lives of fishermen and women. It serves as useful material for the construction of kitchens, yam barns, *atakpas* (meeting places) and houses. *Atakpas* clearly identify Igala residents who use the top part of the building as a yam store and the lower region as a parlour which is in itself a statement about the status of yams there.

Ekwuloku is in Avrugo district of Igalamela/Odolu Local Government Area (Fig. 3.2). What is known as Ekwuloku today did not have a name before 1918. Prior to this the whole area now known as Ekwuloku *gago* was a series of farm settlements run by Igbo farmers from Nimbo that now forms part of the Uzo-Uwani local government area in Nsukka, Enugu state. They named these settlements after their home villages in Igboland. Igalas were actually the second group to migrate to this region; Igbo settlements attracted Igala hunters, farmers and traders. All settled peacefully and intermarriage between the two groups was common. As both these migrations were taking place from Nimbo and Idah the Attah of Igala (the King of the Igala) and the Achadu controlled the whole of the Nsukka area; the Igala kingdom having spread as far as Nsukka. The chiefs of Nimbo, living in Ukpabi, were installed by the Attah at Idah where they obtained their beads of office (their status as chiefs). The Attah controlled all the land and production as people moved freely around the Igala-Igbo border. However, this situation changed when the colonial powers fixed artificial boundaries between Nsukka and Igala divisions; this resulted in Igbos gradually withdrawing from Ekwuloku and Igalas taking over administration of the area. Igbo names were changed slightly so as to have an Igala meaning as Igalas became politically dominant. This was understandable as the Igalas had connections to the Royal clans at Idah. Ekwuloku is one of the villages in Igalaland occupied by a royal clan of

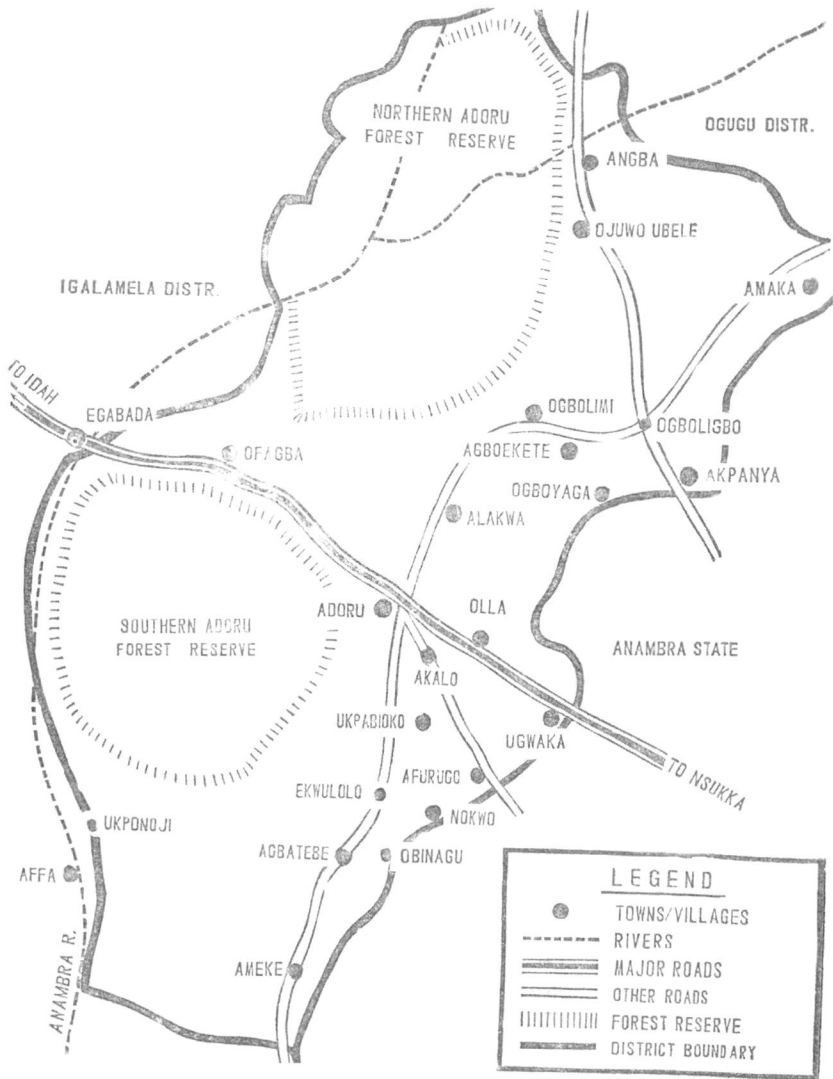

Fig. 3.2 Sketch map of Adoru district

Idah—the Ochai Atta clan—with connections to Angwa in Idah. At present Igbos cannot own land in the village.

After 1918 and with the introduction of the office of *gago* villages in the same area with a similar history were grouped together and given the same name. Ekwuloko already existed as the largest (in terms of population) of the gago villages in an area and hence the collection of gago villages that were nearby was given that name. It is the focal point for all land routes in the region as roads converge there, as well as having both a primary school established in 1976 and a community secondary school established in 2002. At the time of the SLA the village had some 111 households and from a sketch map of the village (Fig. 3.3) it does not appear to be particularly large or indeed special. What makes it especially interesting for this study, and indeed one of the reasons that influenced its choice for the DFID farmer-managed trials, is that it has the largest food market in the Odolu area and a source of seed at times for Igalaland.

Ekwuloko was well connected in the past as its Royal links and powerful chief served it well. During the Nigerian civil war (1968–1970) Igbos left the village (Igalaland was Federal territory) but began to return to Ekwuloku once the war had ended. Igbos constitute almost 70 % of the present population. The current district head of Avrugo attributes the success of the market in Ekwuloko to the return of the Igbos. Without them the market would not stand as they provide the bulk of the yam, cassava and cocoyam. In more recent times Ekwuloku has benefited from other major intervention especially roads constructed by the AADP. The idea was to open up roads to the nearby forest reserves that can be seen in Fig. 3.2 and this literally put Ekwuloku on the map. Although roads generally are poor the construction of the AADP road through Ekwuloku joined it to Avrugo and Odolu and northwards to Akpanya and on to Alloma and Anyigba. Indeed given its history and geography Ekwuloko has become an important political and social centre for the region. As seen earlier, people from here had visited DDS office because of an interest in agroforestry and DDS believed that with its connections with the yam market this would enhance local initiatives for those with an interest in seed yam production.

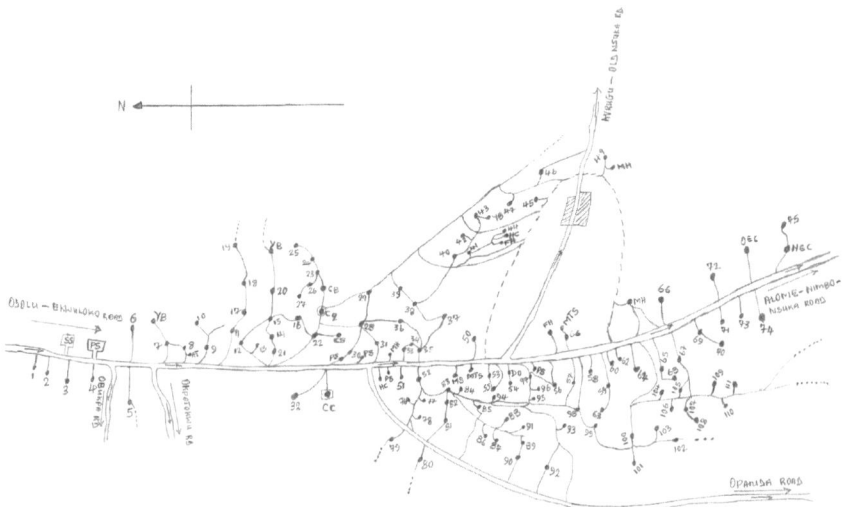

Fig. 3.3 Sketch map of Ekwuloko village. *SS* secondary school, *PS* primary school, *CC* Catholic Church, *CQ* community square, *CS* community shrine, *HC* health clinic, *FH* food hotel, *PSt* provision store, *MS* medicine store, *MTS* market store, *DO* development office, *OEC* Old 'Everyhome' Church, *NEC* New 'Everyhome' Church

3.9 Conclusions

This chapter has set out the broad context within which the SLA discussed in the next chapter took place. Even if the context is broad based it incorporates the micro levels of the national and local societies. It demonstrates the historical tensions existing between the different regions of Nigeria which are in part based upon ethnicity and inter-related at times, with, for example religion. These fault lines exploded during the Civil War and have not disappeared. At the time of writing these have again have become manifest within an upsurge of terrorist violence promulgated by a group called Boko Haram based in the Islamic North fighting to remove the influence of what they see as 'Western' education. It seems they are trying to achieve this by bombing Churches and government offices. There have been reprisals from Christians and many people have died as a result of both the bombings and subsequent revenge attacks. The situation is tragically becoming reminiscent of the events that preceded the Civil War. Other national-scale influences that have permeated down to local livelihoods include the general neglect of agriculture as the oil-based economy grew. In parallel with this diminution of agriculture, of course, was the growth of jobs in the cities and a resulting rural-to-urban migration that denuded many villages of their youth. The HIV/AIDS pandemic also had its toll nationally and Igalaland is not an exception. Policies such as those within SAP have helped redress some of that imbalance but these too have brought their own social problems. It has to be stressed that Edeke and Ekwuloku have not been immune from these larger-scale influences, and what one sees in these villages today have been shaped in part by these forces. Thus while both villages may have little direct government help today, a point which will be returned to in Chap. 4, the livelihoods of the people that live there continue to be influenced by what happens in the rest of the country and beyond.

The SLA described in the next chapter was implemented by a Catholic Church-based agency called DDS, and specifically linked to a DFID funded project designed to explore the potential for clean seed yam production in Nigeria—DDS was one of the partner agencies. It has its own history which began in the immediate aftermath of the Civil War, and was shaped by wider changes occurring in the Catholic Church during the 1960s. Igalaland was a war affected area and suffered much as a result. The decision by the Diocese of Idah to engage in development on a significant scale is easily explainable given this history. The decision of DDS to focus on agriculture and micro-finance is also understandable given the dominance of this sector in the livelihood of most people as well as the presence of indigenous self-help schemes (such as the oja) that formed a natural basis for intervention. During its 40 years plus of existence DDS has worked with many partners; local, national and international. It has been successful in seeking support for numerous projects spanning agriculture, health care, nutrition, education, and rural infrastructure and so on, with much of this support from international development donors. Some of this support included a research component as DDS believes research and development go hand in glove. Hence the engagement of DDS within the DFID yam projects was the latest manifestation of a long history of such engagement and by no means a reflection of a change in direction on the part of DDS. Each intervention is true to expressed needs and wishes with on-going dialogue and discussion with local communities. This respect dominates all action and allows for differing and diverse opinions. No one has a monopoly of understanding and funding can never sway the direction a project takes. Sustainability was also a key aspect of all of this as there was little point in helping to bring about a change if that did not last. The problems that can arise if sustainability is not considered from the onset can readily be seen from what had happened with the AADP but in fairness some interventions like being aware of seed quality still persists. Roads substantially helped change many livelihoods. But sustainability does entail an understanding of the challenges that communities faced and how best to accommodate and address them if at all possible. One also needs to know what resources (capitals) a community has at its disposal and how best to employ them. (Sustainability cannot always be the first consideration as this would exclude many members in communities. A certain level of confidence and resources (capital) are required before delving into sustainability. It is often the second rung in the ladder of development and takes the longest time). Thus if one takes an SLA as a set of principles guiding development interventions then one could argue that the history of DDS is effectively one long-term SLA implemented over four decades but taking into account again the need for diversity as the poorest cannot often make this grade. The formulisation might have not been the same as set out in Chap. 2 as DDS did not catalogue the capitals, shocks and so on, but that is precisely what it did almost instinctively and over time. Indeed this was often implied when making applications for funding. This affirmation illustrates the fundamental logic inherent within an SLA; it does indeed set out all of those aspects that need to be explored before a development intervention can be contemplated. As discussed in the previous chapter the notion of an SLA as a set of guiding principles applied by an agency over the long term may be at odds with the usual application of SLA over relatively short time scales, but it is just as valid. Indeed this helps set the context for why DDS implemented the SLAs in Edeke and Ekwuloko villages; it was seen by them as part of a longer-term strategy rather than being project-specific. DDS knew this would help them develop an evidence-base for planned changes to the microcredit scheme so as to influence its own thinking on this but also to help provide support from external funding sources. DDS therefore took the decision to adopt a more in-depth, intimate and case-study approach to the SLA rather than implement something on a much larger scale across a number of villages. The more targeted approach allowed for a more nuanced and longer-term engagement with a limited number of households and thus their involvement as gatekeepers into the wider community. As far as the authors are aware, this was a unique use of SLA.

   Finally, given DDS's roots within the Catholic Church and its vision of development, it is well to mention that the approach and practice of SLA as analysed in this and the following chapters are derived from what can be called the 'Missionary Way' of being in development. Indeed the recognition of faith—based groups has gained momentum in the past two decades and this at least in part emanates from the Missionary engagement of many faith-based organisations for well over a century in various parts of the world. This approach is holistic in the same way that SLA claims to be, with trust playing a vital role. Belief in the inherent dignity of the human person as the starting point for a moral vision of society is stressed. In a development mix therefore, economics or politics or sociology or psychology or culture alone is not sufficient for holistic development or to form the basis for a sustainable outcome to development at primary group level—the household. In the 'Missionary Way' of development community and the common good are of great import with an obligation to love and respect our neighbour; being our brothers' and sisters' keepers played out in a reciprocal manner. Such an approach is guided by a belief in social inclusion (and hence problems with sustainability as the truly poor are not sustainable) where all have a right to participate in the economic, political and cultural life of the society in which they live. In recent times, the obligation to be stewards of our environment has gained in prominence and with this comes a vision of development striving to bring about long term justice and a wholesome society where sustainable livelihoods can be nurtured and promoted. Negative aspects of our understanding of development can be worked on to create new approaches and this offers challenges to all involved-participants, researchers and development practitioners. In such a context it is possible to envisage various means to end poverty and discord. But livelihood is just part of the story, and also of importance are people's wellbeing (including spiritual) and lifestyle (their way of living), especially their expectations. Development is not just a matter of reducing poverty, and livelihood is but part of the picture. While faith-based groups are just a part of the much bigger development jigsaw, this wider vision as to what development entails is something that will be returned to in a later chapter as it has repercussions for SLA.

Chapter 4 will take the story one step forward by describing the findings that emerged out of the SLA's in Edeke and Ekwuloko. To some extent these affirmed what was already known about the two villages and set out above, but there were also surprises.

Chapter 4
The Sustainable Livelihood Approach in Practice

4.1 Introduction

The previous chapter set out the context for the SLA and the reasons why DDS believed it would be beneficial. There were a number of interacting factors at play in the decision, including:

- The involvement of DDS in a DFID funded project on seed yam which included a series of farmer-managed on-farm trials with new methods of producing healthy seed yams
- A desire to rejuvenate and revamp the micro-finance scheme that DDS had been operating since its inception in the 1970s
- The need to include new areas of Igalaland where DDS had only a brief presence (primarily the Igala-Igbo border)
- Prior experience with numerous surveys in Igalaland
- A perceived need to ensure that potential donors are convinced of the value of supporting the new micro-finance project

The combination of these factors facilitated the decision by DDS to 'piggy-back' an SLA onto the farmer-managed trials with clean seed yam that were being funded by DFID. This would provide a better background for appreciating the impact that would result from the healthy seed yam technology, and be beneficial for the DFID project. It also gave DDS scope to explore the potential for its new credit scheme (also of relevance to the DFID project) while investigating opportunities for including new areas. Funding was provided in part by the DFID project, although this was inevitably limited given the primary focus on the farmer-managed trials. Other agencies (e.g. Gorta) assisted in the study as they were interested in the refocusing on the credit scheme for food security.

At first glance the emphasis on only eight households in the two villages (four per village) might be surprising. After all, how can these households be in any way representative of the wider populations of those villages, let alone Igalaland? Given that DDS was in the throes of bringing in a new microcredit scheme for its

S. Morse and N. McNamara, *Sustainable Livelihood Approach*,
DOI: 10.1007/978-94-007-6268-8_4,
© Springer Science+Business Media Dordrecht 2013

mandated area, then this might at first glance appear to be an odd decision. Why not implement a more wide-ranging SLA across a larger sample of villages? Yet Chap. 3 has presented the counter-argument to this and has set out the logic as seen by DDS.

In this chapter the focus will be what DDS did in the SLA and some of the main findings that emerged from the process. It will also seek to explain what DDS felt it has learnt from the process and how this helped them in influencing its planned changes to the microcredit programme. An important consideration throughout is the logistics of the SLA process. The main findings from the SLA are summarised in this chapter using the structure of the process set out in Chap. 2 and Fig. 2.1. The chapter will begin with a brief discussion of methodology followed by the results of the SLA and the challenges that had to be faced. Each section will remind the reader of the components being addressed.

4.2 The Sample Households

DDS was well aware from previous experience of working in Igalaland that SLAs are demanding in terms of resources, and this was a major factor in its decision to focus upon only two villages. However, even with this limitation, the populations of Ekwuloko and Edeke are substantial, each comprising hundreds of households (herein referred to as HH). It was simply not possible to include all of them within an 'in depth' exploration of livelihood and a decision was made to focus on just four HH in each village. The size of an optimal 'sample' in SLA is a moot point and much depends upon what is meant by 'optimal'. To some extent this implies a sample that is representative of the population, but there is clearly a trade-off given that larger samples can typically equate to higher cost. In addition there is a third element to consider, namely quality, as any information collected has to be a 'real' reflection of livelihood and not a construct. These three demands of SLA can be presented as three overlapping circles with SLA being the point of overlap (Fig. 4.1). The three circles 'pull' against each other, such that there is almost always a tension to reduce cost and maximise quality and representation. Hence the demands of the process create pressure to use smaller sample sizes for best quality while the competing demand for 'representation' demands larger samples. But if the cost circle is over emphasised then the effect will be to diminish representation and indeed quality. Similarly if the emphasis is too much upon representation with larger sample sizes or indeed the whole population then the effect may be to dramatically increase cost and reduce the quality of the data.

The decision to focus on eight HHs over the two DDS SLAs was inevitably a compromise and one that can readily be criticised. An alternative would have been to rely more heavily on observation along with a larger sample and rely upon less intensive methods to generate information that could be put together to formulate a picture of livelihoods. This would have had the additional advantage of picking up on the variation in the sample as people have diverse livelihoods. But this would have been at a price as there would be a substantial reliance upon

Fig. 4.1 The three demands
of an SLA. Note how the
demands of the 3 circles
can 'pull' against the others.
For example, there may
be pressure to reduce cost
(the *arrow*—resulting in
smaller circles) and this
inevitably could also reduce
representation and quality
such that the overlap—
SLA—becomes untenable.
Even if the reduced cost
results in less representation
rather than quality the point
of overlap could disappear

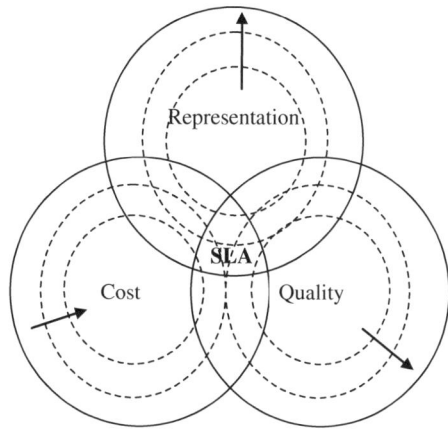

respondents telling the truth as it would be impossible to verify all the responses
with observation. There is a real danger of putting much effort into collecting information which may at the end have a limited basis in reality. DDS had
limited resources at its disposal so inevitably there were trade-offs which had
to be made. That said, the early phase of the DFID project (conducted in 2003)
included a broader survey of households in a number of Igalaland villages, including Ekwuloko and the Edeke area. The aim of that phase was to explore problems
with yam production, and included a series of focus group meetings with farmers
followed by use of a structured questionnaire-based survey which built upon the
main findings of the focus group phase. While the survey was focussed on yam
production rather than HH livelihood, it did ask questions about credit availability
and expenses incurred using local credit facilities; information was sought on markets used by farmers for selling their produce. These survey results helped identify
Ekwuloko and Edeke villages as suitable sites for the on-farm trials.

A second consideration regarding methods was the piggy-backing of the SLA
onto the DFID funded seed yam project. The project included a number of on-
farm (farmer managed) trials in both Edeke and Ekwuloko to test the technology and present it to the farmers for their evaluation. This entailed the selection
of a small number of households in each of the two villages and DDS saw this
as a tempting opportunity to work with those same households for the SLA. It
would help to reduce the cost of the SLA as regular field visits (funded by DFID)
had to take place for the on-farm trials and the continued engagement with the
HH would also help build trust. This was not a one-way process. There were, of
course, overlapping benefits here as while the detailed SLA may not have been
'formally' required for the DFID-funded project, the insights gained were useful
in that they helped set the gains from the seed yam project into a wider context of
livelihood. Also given that the promotion of clean seed yam by DDS was going
to be linked to its micro-finance programme, then any information which helped
with the reformulation of that programme would be of benefit. It should be noted
that a key consideration for DDS was the potential for reorganising its microcredit

scheme focusing more on a 'business plan' approach and promotion of clean seed yam enterprises would be a integral part of this. There were two aspects to this:

1. Need for evidence demonstrating how a business plan and enterprise approach could be made to work in practice
2. Creation of credibility with potential donors who could provide support for the transition.

Thirdly DDS wished to work closely with a relatively small number of households as they were seen as possible 'gate keepers' into the wider community, especially in Ekwuloko where DDS had little prior presence. A DDS presence had existed in Edeke for many years so this was less of an issue there; but Ekwuloko was new territory and not as familiar to DDS as its border location between Igala and Igbo on a bad road made it out of reach. In other ways the Ekwuloko area was well known to DDS, especially for its yam markets where DDS staff went to buy seed yam. However, DDS felt there was an absence of 'gatekeepers'; prominent members of the society who could provide advice and help to build trust with their local communities. DDS had operated through trusted local gatekeepers even since its inception and found this to be an excellent approach in the hierarchical Igala society. In the early days the gatekeepers were expatriate, often Canadian priests, but there was a snowball effect as prominent Igalas also became involved in DDS projects. It was only natural for them to think the same way with regard to exploring an expansion into Ekwuloko.

Once DDS made the decision to piggy back the SLA onto the DFID-funded seed yam experiments the next decision was the selection of the participating households. Village meetings were convened in Ekwuloko and Edeke to help select the four HHs in each place and to explain what DDS was trying to achieve with its reorganisation of the micro-finance scheme. A key issue, and one that is always something of a conundrum for DDS, relates to the hierarchical nature of Igala society. This demands that the 'seniors' lead while others follow. This can be advantageous as it can allow for the use of the 'gatekeeper' model highlighted earlier but much depends upon these individuals concerned. In Ekwuloko this meant the programme had to include the village chief (gago) and of course the most senior Igbo given its historical context. As the Ekwuloko SLA could only include four participating HHs, this immediately limited choices for the remaining two. Out of deference to Ekwuloko's ethnic composition another Igala and Igbo HH were selected in cooperation with the villagers. The relative unfamiliarity with DDS was at times problematic but this was compensated by the seniority of the four HH heads—a factor which offered an opportunity for them to be potential gatekeepers into the wider community. No selection problems occurred in Edeke, largely because the village had a longer and deeper interaction with DDS; Edeke people were more familiar with how DDS worked and the nature of its credit scheme. The four HHs were selected based on their being long-standing members of the DDS saving and credit scheme; they were also regarded as 'opinion leaders' in the community. Here the opportunity as perceived by DDS was more about cementing an existing relationship.

The SLA in Ekwuloko and Edeke was based on similar work already undertaken by DDS in other villages in Igalaland since the 1970s and would of course concentrate on HHs as the social unit. SLA was a familiar concept to DDS and its staff had experience, albeit in other Igala villages rather than Edeke and Ekwoloko (Morse et al. 2000). A student studying for a master's degree undertook research for an ethnographic dissertation in Edeke in 1999 and this provided some useful background, albeit somewhat dated. Despite tangible differences between Edeke and Ekwuloko and the many villages in which DDS worked, the same working definition of a HH was adopted as that in Morse et al. (2000):

> A HH is a clearly distinguishable social unit under the management of a household head (HHH). The HH shares a community of life in that they are answerable to the same HHH and share a common source of food.

In the past a HHH in Igalaland was usually male but this is changing reflecting conditions in Nigeria (and many other countries) caused mainly by male migration to urban centres and the HIV/AIDS pandemic. Consequently female HHH are far more common than in the past. Although a HH may typically comprise blood relatives, others including in-laws are often part of it. HHs are therefore 'volatile' in the sense that size can vary substantially during the year and between years, and include friends as well as relatives. A HH typically occupies an area with a collection of buildings for habitation and storage, a unit referred to in Nigeria as a 'compound'.

With the definition of the concept of HHs agreed upon, the steps involved in the SLA were broadly as follows:

- Assessment of human and social capital (household composition, education and skills, societal membership)
- Assessment of natural capital (land, farming, tree crops)
- Assessment of physical capital (machinery, tools, buildings)
- Assessment of financial capital (income, expenditure, savings)
- Resilience of these assets to change and institutional context

While these are listed in sequence they did not necessarily take place in that order. Thus, for example, an exploration of resilience and institutions was pursued throughout the year. Also, to some extent the process was guided by knowledge which DDS had of the two villages. The existing land use by the HHs, their physical assets, income/expenditure, human and social capital were included in order to appreciate what livelihood options were available to the HHs.

The SLA began in early 2004 (Ekwuloko) and early 2005 (Edeke) continued to early 2005 and 2006 respectively. Each SLA lasted for approximately one year, and in each case ran in parallel with the DFID funded seed yam trials. While much of the data was qualitative obtained through observation of HH activities and informal discussions, there was also a considerable amount of quantitative data collected during the year. This rich dataset will be briefly summarised in this chapter. As DDS has often experienced, there was a high level of cooperation (which happens gradually) from all the villagers and not just the eight HHs. This required putting energy into the establishment of trust and carrying out activities immediately of benefit to the other village dwellers. Once villagers realised DDS was not seeking data that put ownership of their land in jeopardy and was not reporting to some government institution for tax purposes, then data collection became far easier.

The methods employed in the SLA included:

– Formal interviews based on semi-structured questionnaires
– Informal discussions (individual and group)
– Field mapping (primarily the land owned/rented by the HH, but also buildings)
– Observation and participation in activities

These are well-established methods in SLA and are familiar to DDS staff and presented no difficulty other than the sheer quantity of information that needed to be collected. A number of DDS staff took part in the process, and while the local dialect of Igala spoken in Ekwuloko was noticeably different to Igala spoken elsewhere, it was not a hindrance to communication. The work involved regular visits to the villages and necessitated DDS staff having to reside there for some time (usually with the HH). Results were routinely checked with Igala key informants living outside the village, and two particularly useful informants were the District Head of Odolu and his wife who hails from Ekwuloko. DDS found the use of key informants invaluable, especially as they could help with the interpretation of findings that might be quite different when compared with findings from other Igala villages.

4.3 Human Capital: The Households

Human capital (highlighted in the context of SLA as Fig. 4.2) provides labour for the various enterprises (income generation, subsistence farming, water collection etc.) engaged in by the HH. While human capital is partly related to HH size, much also depends upon levels of education, experience, age, gender profiles, and

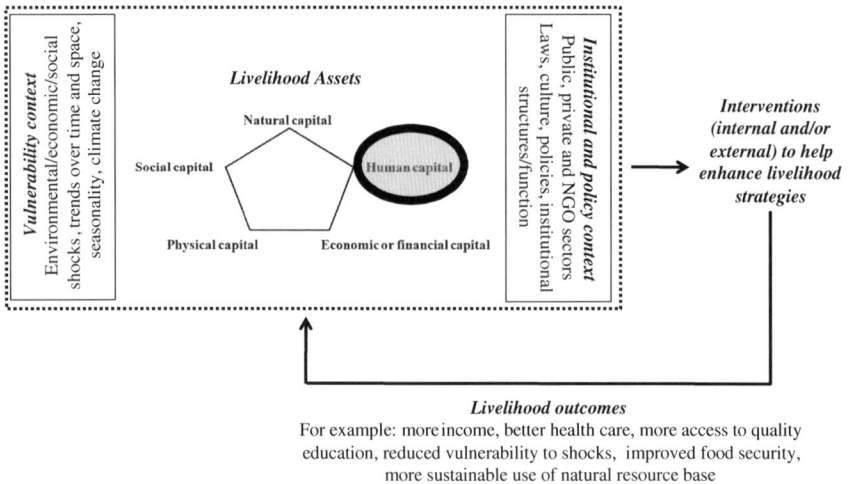

Fig. 4.2 Human capital

occupations and so on. Many activities are 'gendered' in Igalaland (i.e. predominantly carried out by either men or women) and children help with some tasks.

Figure 4.3 is a snapshot of the composition of the selected HH in Ekwuloko and Edeke as it was at the time of the SLA. The household codes (M1 to M4 for Ekwuloko and E1 to E4 for Edeke) were used to preserve anonymity. As already mentioned, Igala HH tend to be fluid in terms of people arriving and leaving but fortunately they were more or less constant over the years for the selected HH. However, there were complications. M3 in particular was a 'bi-locational' household as this HH had a base in Ekwuloko and Nsukka. Thus some HH members

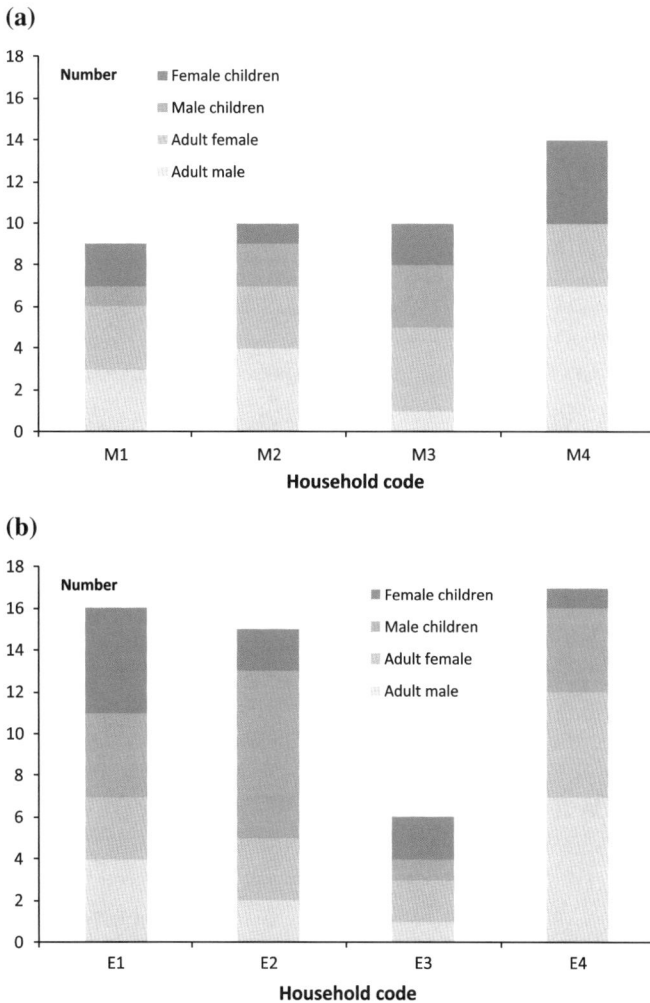

(a)

(b)

Fig. 4.3   Composition of the selected HH in Ekwuloko and Edeke. **a** Ekwuloko. **b** Edeke

were divided between these two resulting in some exchange in location, but for the sake of simplicity they are all recorded here as members of the Ekwuloko HH. By 2011 HH M3 had left Ekwuloko.

The four chosen HHs in both Ekwuloko and Edeke differed in size, particularly in the number of male adults. The term 'adult' as commonly defined in Nigeria is anyone aged 15 and above while a 'child' is anyone aged less than 15. One of the Edeke HH (E3) was the smallest of all eight HH surveyed in both Edeke and Ekwuloko, comprising six people and only three of these were adults (one aged 89 years). HH size does matter for livelihood sustainability as the number of adults is an important determinant of labour availability for income generation, but is not the only source as labour can be hired.

One point that immediately became apparent from the composition of the HHs is the differing levels of education amongst members. Such variation in levels of education is not unusual for Igalaland, and has been summarised in Fig. 4.4 (figures refer to the highest level of education achieved by a HH member). Education is important as it can help with off-farm sources of income; more educated members have a better opportunity to earn a wage or salary. HH M2 had a particularly high level of education, with four members educated to tertiary level. The HH Head was highly educated for his generation and obviously believed in the same for his children. Despite HH M4 having no formal education, all his children of school going age were in school at the time of the study though he admitted they began school relatively late. This was because they had helped him with household chores. He did not expect them to follow in his footsteps in farming given their level of education but like many more in Igala he hoped they would return to the land when they retired. According to HH M1 farming was for those who had no formal education and his children too had more formal education than he had. HH E4 also had a particularly high level of education relative to the other three HHs in Edeke, with almost eight of the HH members having had at least secondary education. In common with much of Igalaland there has been an increase in levels of education in both villages over the last 20 years almost always accompanied by an expectation that education results in migration to centres of employment (i.e. the major cities in Nigeria).

The following is a brief summary of each HH and its head, and in particular how they appeared to the DDS staff involved in the SLA.

4.3.1 Household M1 (Headed by the Village Chief)

Even if the selection of the Igala chief or *gago* of Ekwuloko was inevitable for the DFID seed yam trial (and hence the SLA), it was to a certain extent serendipitous. He was educated, keen on all interventions suitable to his needs, a full time farmer, and also generous in donating land to those who needed it for farming (although they did pay rent). The main crop cultivated by him was yam, another reasons for his selection. He had encouraged others interested in agriculture to

(a)

(b)

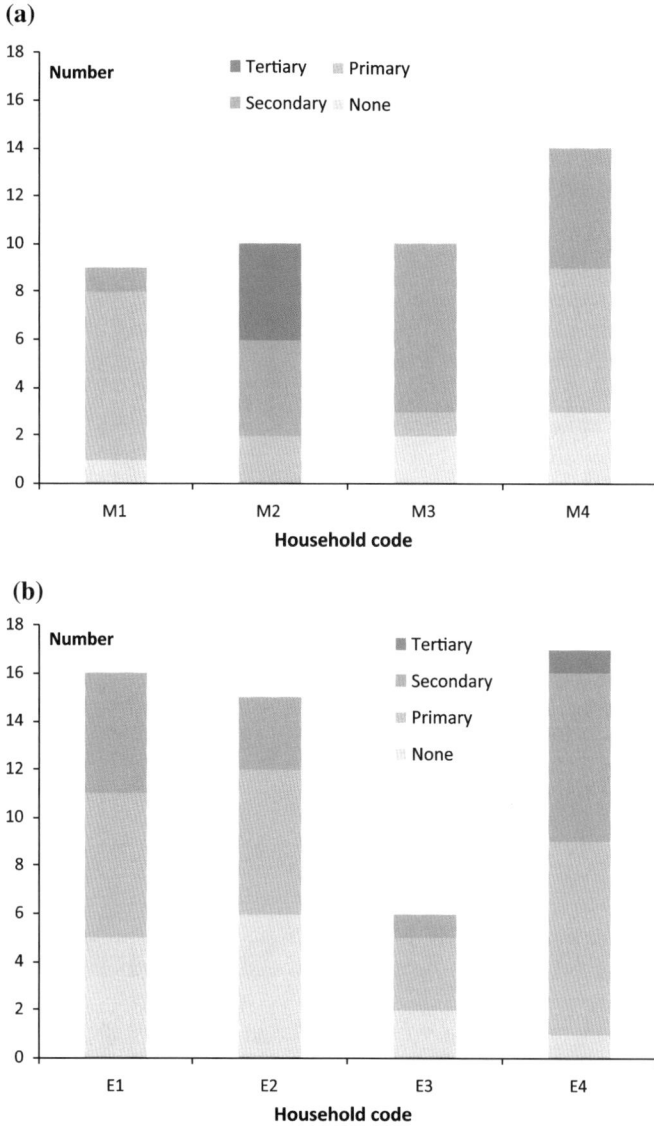

Fig. 4.4 Education of the members of the selected HH in Ekwuloko and Edeke. **a** Ekwuloko.
b Edeke

do likewise and his concern to train local young men in skills that provided extra
income to supplement on- and off—farm activities seemed genuine. His interest in trees was well compensated by the income from them; green legume cover
crops also appealed to him. Overall, M1 had skills that helped in the unification of
the community; a positive attribute wherever it is found but especially in a local

leader. However, with use of the key indicators of social capital to measure wealth it was perhaps surprising that the local chief emerged as the poorest of the four Ekwuloko HHs.

4.3.2 Household M2 (Headed by a Senior Igbo)

The second HH Head selected for the SLA was an Igbo, a part time farmer and full time civil servant. He was an informal local leader well versed in both cultures who because of his profession as a school inspector spent some time travelling in the locality and to Idah. Of all the four he was perhaps the best acquainted with DDS in that he had heard of its work in other villages and visited the office for advice. Indeed he regretted not being able to avail of DDS projects in the past because of a lack of finance. In common with the three others he ran an almost model farm and seemed to be a natural at recommended agricultural practices. Of all four HHs, he was the most highly educated (to third level). This combination of good education and the respect shown him by the villagers were the main reason why they suggested that he be included in the DFID seed yam project and hence the SLA.

4.3.3 Household M3 (Igbo Community Leader)

The third person chosen had previously been registered as a FC member but in another village. He was both a politician and a community leader which is not too unusual in Nigeria and this explained why he was put forward by the community. His interest, ability and skills in agriculture were obvious but he was an entrepreneur in the broader sense of having both on and off-farm enterprises within Ekwuloko and in his native village close to Nsukka (Igboland). He was an Igbo, an immigrant and rented land in Ekwoloko. He was proud of being totally assimilated into all socio-cultural and political activities in the village. At the time of the SLA he had two educated wives both in paid employment one took charge of the HH affairs in Nsukka where some of the children were in school. It was obvious he lacked capital since some of his resources, for example a palm nut cracker, could not be utilized, as he needed money to commission it.

4.3.4 Household M4 (farmer and business man)

The fourth HHH selected in Ekwuloko had no formal education but again with the use of the indicators employed in this research, he was likely to be the wealthiest of the four. He was a driver by profession, a job that brought him to all parts of the country—north, south, east and west—and indeed well beyond the borders of Nigeria. Like many of his Igala counterparts who also travelled to other African countries he came back with skills, equipment and materials that he in turn adapted to local conditions. As he put it, he "did not savvy book—he savvy the things inside the book" (meaning that he had little education but has much knowledge of life). He was a born entrepreneur. An Igala with an Igbo first name and two Igbo wives who were much junior to him, he was hopeful his sons would eventually succeed him in farming. He admitted to having enjoyed the "better life" (referred to in local parlance as "playing guy") for many years before returning to Ekwuloko. As a multi-skilled worker and widely travelled, he seemed able to adapt to all situations achieving the highest of standards. He farmed extensively with a high standard of husbandry. His method of organising his farm labour force was unusual and innovative for Igalaland in that he had families staying with him ("friends") as lodgers who provided their labour as payment. This accounted for the relatively large number (7) of adult males in his HH where everybody appeared happy with the arrangements. The HHH of M4 was ready for any intervention that would be to his advantage as he had the capacity to investigate all angles of a situation. Everyone in the community would watch him carefully and this expertise explains why he was selected by them to take part in the seed yam trials.

4.3.5 Household E1 (Farmer and Vigilante)

HHH E1 in Edeke was a person who combined farming, fishing and hunting in that order of importance to him. He was the only one of the four Edeke HHH selected for the SLA who was not originally from Edeke but from the nearby local government area of Ibaji. He also headed the local vigilante groups, which earned him income from the local council. Crime was something of a growing problem in Idah local government area at the time of the project and it was not unusual for areas to establish their own vigilante. As the vigilante chairman, prominent people from outside Edeke donated money to him because of his ability to mobilise village groups to protect Edeke from crime and maintain peace. Politicians also offered cash assistance for peace keeping during their political tours. He assisted the *Madaki* in settling local cases, which no doubt earned him some 'tokens'. As a landowner in areas outside Edeke, he took part in sharing farmland and arranging rent. His wives were into farming, trading and other income generating businesses. At the time of the research many of his children were in schools and still readily available for farm labour. Fishing was vital to his income generating business and his wives purchased part of the fish harvested by both children and husband and sold them in local markets. Sales from farm produce covered all household requirements but the money made from sales did not cover costs of planting materials for the following planting season. Farmers began each planting season with a debt given the high cost of planting materials; indeed local debt was a crippling factor as farmers lived in a state of perpetual indebtedness. Moneylenders there were not the sharks many believed them to be in other places. Rather they often lost considerable capital and were regularly the victims of bad debt. The family were Muslim with wives and children belonging to the one Muslim denomination in their village of Ojigagala-Edeke and Idah. He had three wives, six sons, four daughters a son in law and daughter in law all of whom appeared to be living happily together. He spent the evening line fishing. He gave part of his catch to his friends who did not have the opportunity to fish. There was no reciprocity here as beneficiaries are mainly widows and older women.

4.3.6 Household E2(Madaki of Edeke)

An Igala from an Idah clan who heads this household was born in Edeke. He had three wives and numerous children. His education *lasted* to primary five after which he had to assume family responsibilities. He married while in primary school, as he was older when he commenced primary education. His first wife, who was illiterate, was the same age as himself. His second wife was literate as were all his children. His third wife was also literate. They too like household E1 embraced the Islamic faith as had the family for generations. He was first a farmer, fisherman and herbalist. The three wives farmed extensively and the second wife engaged in the fish industry. The household head was also the Madaki (Chief) of the village with responsibility for a number of clans. None of his children were old enough to be employed so there were many expenses in this HH. He belonged to a number of social groups in the village and this membership enabled him to benefit from group labour, credit facilities in cash and kind. His second wife also assisted him with food ingredients as she traded in a number of different locations thereby securing bargains that helped the household economy.

4.3.7 Household E3 (Farmer and Fisherman)

The household head of E3 was the youngest and the most educated of the four HHH. His ancestry could be traced to an Idah clan that migrated to Edeke almost 400 years ago. On completing his secondary education in Idah he obtained a white-collar job in Kaduna. He remained in Kaduna for five years after which time he returned home as his father had died and his mother was old. His only option was farming for which he had to borrow money to get started. His choice was yam production. He engaged in tree crop production taking a lead role in initiating tree crop maintenance in the locality. He also prided himself on being a skilled fisherman. To take care of his responsibilities effectively he engaged in hired labour for which he was paid. He also received help from the *Ayilo* group (rotational labour organisation) of which he was a member. At the time of the SLA he participated in ten other social groups, mainly in agriculture. He was registered as a member of the ruling party in Nigeria, namely the People's Democratic Party (PDP). His only wife was literate and not from Edeke or Idah. A friend who had married from the same village as his wife introduced him to her. His children of school going age attended the local primary school. The youngest was only a year old. He took care of his aged mother who was now blind. He had been secretary to the DDS micro-credit programme in Edeke since 1998 and applauded by the members for the work he had done. His interest in yam production and long experience with DDS were the main reasons the community appointed him for the DFID seed yam project.

4.3.8 Household E4 (Madaki in Edeke)

This household head was the most senior of the four respondents. He too had *Madaki* (chief) status in his village and held the post since 1992. His jurisdiction extended to seven clans. He attended adult education classes 15 years ago and obtained a certificate following his three year course in Edeke. He belonged to many social groups and assumed lead roles in some of them. At the time of the SLA he chaired the *Achekaje* yam producer's association (Idah branch). Responsibilities embraced such matters as market research; ensuring yams were sold where prices were best. This association ensured that no particular group had a monopoly over yam purchase from local yam producers. They also regulated supply to local markets so that there was never a glut that would cause prices to fall. They also settled cases related to loan defaulters especially in relation to yam. Because of his long-term proficiency in yam production he had earned a special Igala title that means he had produced a certain number of seed yam varieties that completely filled a barn. This title is known as *Egwele (Onwa)*, highly coveted in Igalaland, and the meeting of seed yam producers is called *Ujeju Onwa*. Only farmers who have reached a certain level of proficiency are admitted to the *Ujeju Onwa* meetings the *Egwele* attends. He had long been a member of the *Ayilo* group from where he had enjoyed labour benefits on many occasions.

He prided himself in farming but had other strings to his career connected with agriculture. Foremost among these were his profound knowledge of herbal treatments which made him custodian of many secrets and much indigenous knowledge. The younger members of his family, both female and male, worked with him in the herbal business and became privy to many of his secrets. His renown as a herbalist had resulted in him being engaged as a consultant nation wide. His father, who was an expert in his time, passed on the gift to him. He claims he has inherited extra power in hunting and fishing and was keen that his family inherit this legacy. A skilled hunter and fisherman he supplemented his farm income from these activities. As he had five wives he had many children all of whom had been to primary and secondary school as well as some who had reached tertiary level. At the time of the SLA a number of his children were in gainful employment but each of them, especially the men, had an attachment to agriculture, herbal treatments, fishing and hunting.

4.4 Natural Capital: Land and Farming

Since all Igala households are engaged with agriculture to some extent then land is obviously an important natural resource in both Ekwuloko and Edeke. Indeed any analysis of livelihood in Igalaland has to begin with this, so it is no surprise that

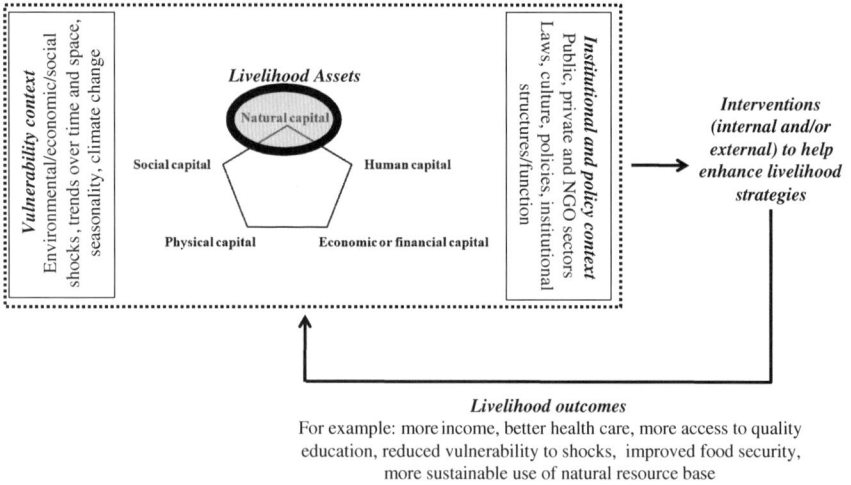

Fig. 4.5 Natural capital (land and agriculture)

(a)

M1

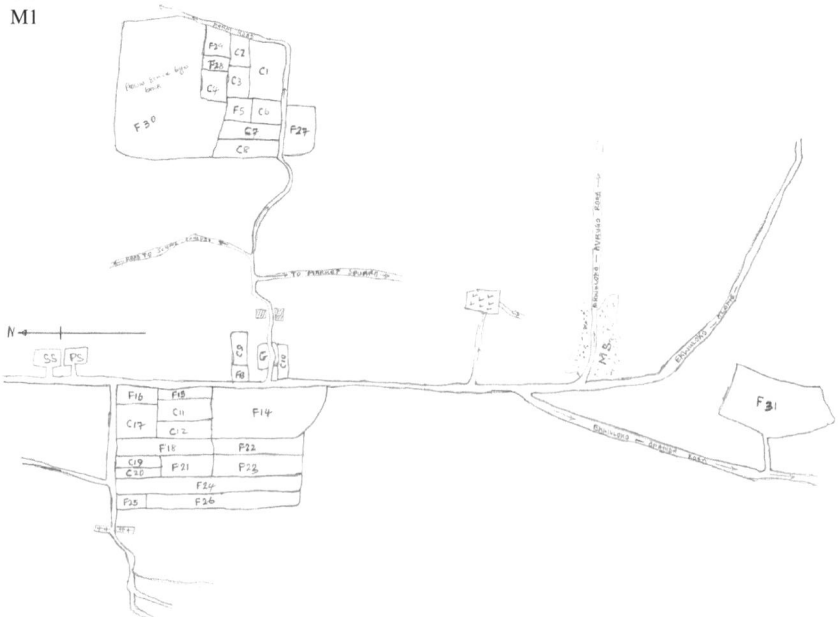

(b)

M2

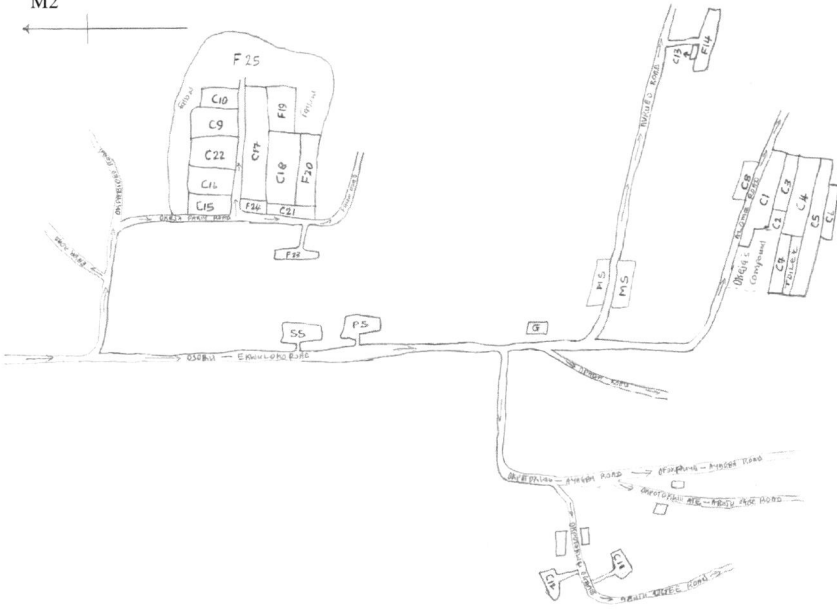

(c)

M3

agriculture and land were important foci of the SLAs implemented by DDS. Based upon a field mapping exercise the plots which the Ekwoloko HH have access to is given in Fig. 4.6. The spatial distribution of plots tends to be clustered. Plots may be spatially grouped but there can also be some significant separation. There is typically a cluster of plots adjacent to the compound but there are also significant holdings some kilometres away. Distributions of land to which HH have access, such as those in Fig. 4.6, are common in Igalaland and indeed throughout West Africa. Given the distances that can be involved, there are clearly logistical concerns such as travelling to and from the plots and these can also be relatively inaccessible to motor vehicles which can hinder transportation of farm produce after harvest.

The areas of land available for the Ekwuloko and Edeke HH are shown in Fig. 4.7. In Ekwuloko the two Igala HHs (M1 and M4) had the largest land area, the greatest number of plots and they owned their land. By way of contrast, the two Igbo HHs (M2 and M3) had smaller areas of land and fewer plots, although this is more noticeable for M3, and both of them rented land. This pattern is not surprising given that Ekwuloko is an Igala village and the Igbo HH are, in effect, immigrants and thus unable to own land. What is perhaps surprising is that the area farmed by M2 is quite close in size to M1 who actually owned his land. M4 was substantially ahead of the other three farmers in terms of land ownership.

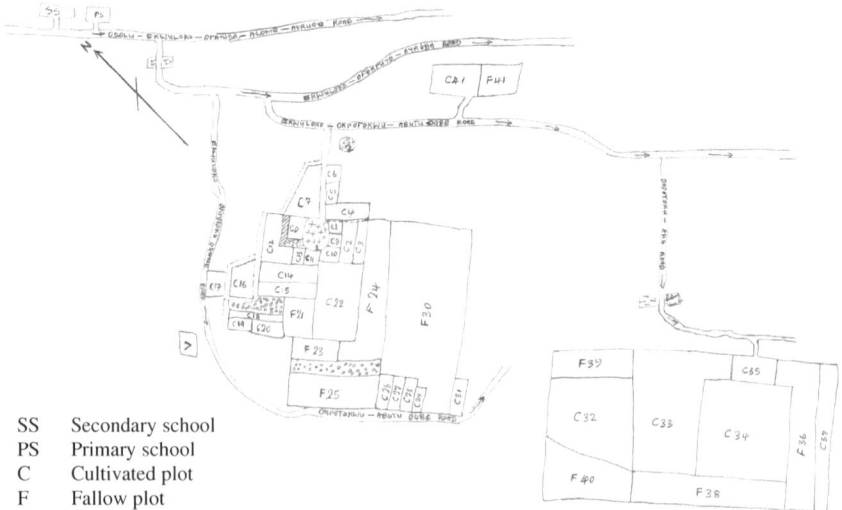

SS   Secondary school
PS   Primary school
C   Cultivated plot
F   Fallow plot

Fig. 4.6 Plot distribution for the Ekwuloko households. **a** M1. **b** M2. **c** M3. **d** M4

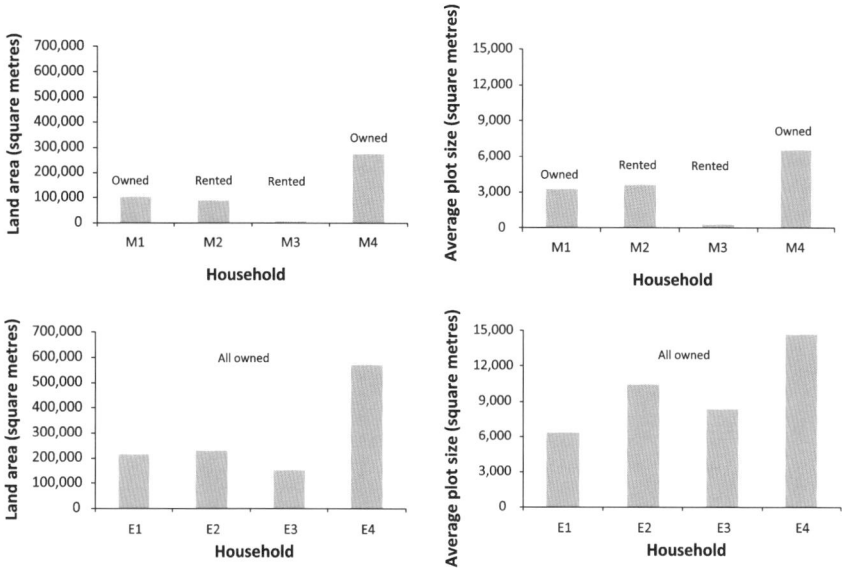

Fig. 4.7   Land areas available to the HH in Ekwuloko and Edeke

Average plot sizes for M1 and M2 were similar. The average plot size for M3 was small, less than 500 m², and while that for M4 was 1.5 ha. The selection of these four farmers certainly introduced a degree of diversity especially for land ownership which is linked with ethnicity. In contrast to the Ekwuloko HHs, all four Edeke HH, were Igala and they claimed to 'own' their land while in fact they paid a tribute for it and thus cannot be said to 'own' their land in the same way that the Igala HH in Ekwuloko did. This is linked to the pattern of land owner- ship in Edeke as, in fact, most of the HHs are of migrant origins. Also, while the number of plots per farmer (18–39) for Edeke were similar to those of Ekwuloko the total land area (cultivated and fallow) resulted in the average plots size being much higher in Edeke than in Ekwuloko. The only Ekwuloko farmer comparable to the Edeke respondents was M4. Thus in terms of land there are marked differ- ences between the two villages and within them, most noticeably for Ekwuloko. The next question is: for what purposes did the farmers use the land that they had access to?

Both Ekwuloko and Edeke are rural and thus it is unsurprising that agriculture dominated land use in both villages. Farmers in Igalaland are almost entirely ara- ble; there is little, if any, pastoralism except for the nomadic Fulani. The farm- ers in Ekwuloko and Edeke are no different in this regard, but cultivation patterns were different between the HH farms of the two villages (Fig. 4.8) as well as var- ying within each village. In Ekwuloko, the two Igala HHs (M1 and M4) had a relatively large proportion of their land under fallow at any given time. For M4 this proportion was approximately 80 % while for M1 it was nearer 90 %. These

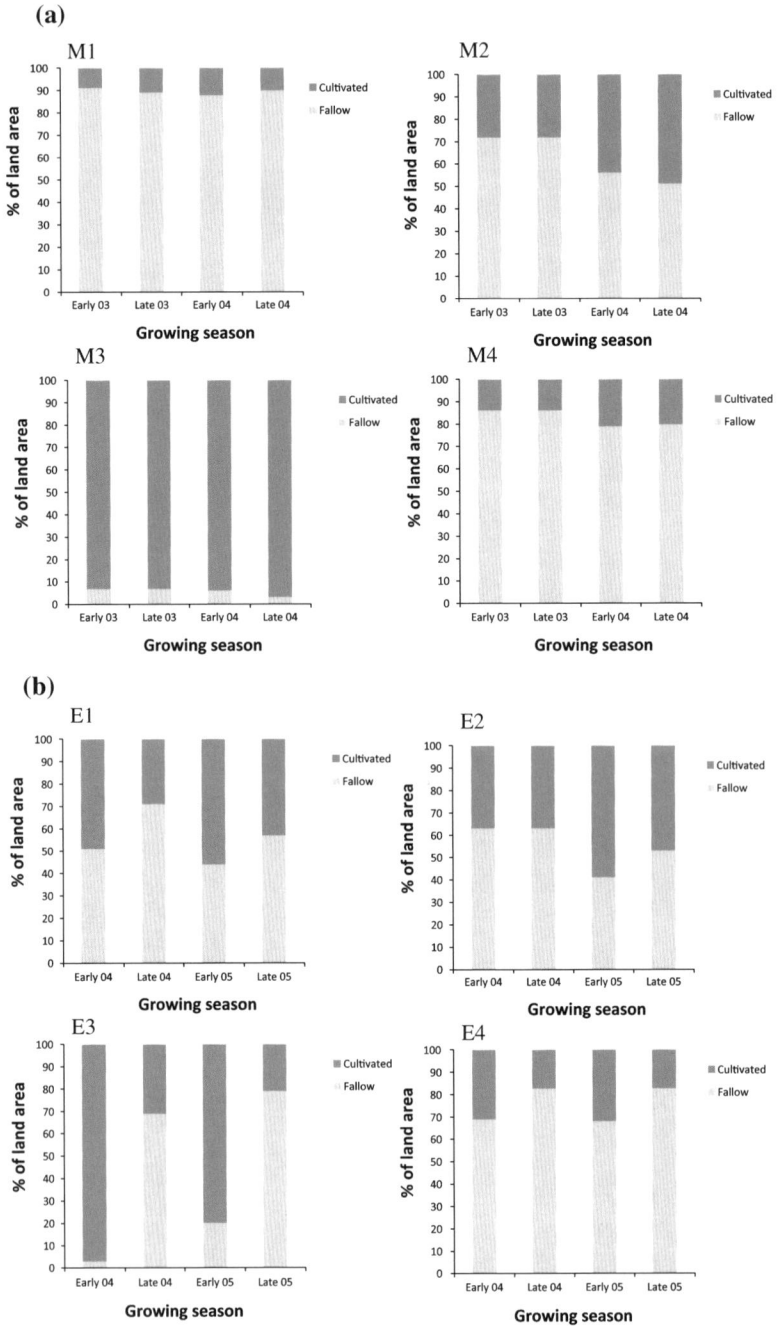

Fig. 4.8   Fallow and cultivated areas. a Ekwuloko. b Edeke

proportions were in fact relatively high compared with other areas of Igalaland, and suggested that at the time of the SLAs land shortage had yet to bite as has happened elsewhere. Indeed the relatively low pressure on land helps explain why Igbos migrated to Ekwuloko to farm and were welcomed. For the two Igbo HH in Ekwuloko the proportion of land under fallow tended to be much less, and for M3 it was less than 10 %. No discernible pattern in the area under fallow could be seen for any of the Ekwuloko HH across the two years—2003 and 2004—or between the two growing seasons in each year (early season = April to August, and late season = August to October). For the four Edeke respondents the intensity of cultivation was higher than for the two Igala farmers in the Ekwuloko sample, although there were differences between the early and late seasons. Land in Edeke is more intensely cultivated in the early season compared to the late season, a fact readily explained by the flooding during the latter part of the rainy season. Thus farmers have to get their crops planted and harvested early.

The crops which were grown by the HH in the two villages are shown in Table 4.1. The pattern was quite different between the two villages but relatively consistent between the HH within each village. The diversity of crops grown by the four Ekwuloko respondents was significantly higher than for those of the Edeke respondents, and the difference can be attributed largely to the production of grain legumes (cowpea, groundnut and pigeon pea) and some vegetables (melon and okra) in Ekwuloko and their notable absence in Edeke. The riverine soils of Edeke are totally unsuited to legumes and many vegetables. There were variations in the other crop categories as well but these tended to be a result of substitution; cocoyam in Ekwuloko for sweet potato in Edeke and the same for guinea corn and rice. Leguminous crops 'fix' atmospheric nitrogen and thus their presence usually suggests that farmers are using them to help manage soil fertility. However, while the absence of grain legumes in Edeke may seem to be a negative factor in sustainability terms, it should be remembered that Edeke does benefit from an annual flood which replenishes plant nutrients and helps control soil-born pests and diseases. Thus not only are leguminous crops not suitable for the land at Edeke they are not necessary.

In common with many places in Igalaland and indeed West Africa, cropping in Ekwuloko and Edeke tends to be dominated by intercropping; growing of more than one crop on the same piece of land at the same time. Intercropping is popular in the local context largely because it provides an element of insurance if one or more of the crops fails due to environmental reasons such as drought or pest/disease attack. Thus all the effort expended in clearing and land preparations are not wasted as at least one crop may survive. But at the same time intercropping is an intensification of resource use; more crops on the same piece of land does result in a more intensive use of resources such as light, water and soil nutrients. Thus the extent of intercropping can be an indicator of many factors such as pressure on land, volatile markets, environmental variation, relative shortage of labour or simply the crops grown as some will not do well under intercropping. The proportion of cropped area under intercropping systems for the HH is shown in Fig. 4.9. During 2003 and 2004 in Ekwuloko much of the cultivated area of the four farmers was intercropped, although there were exceptions as some cassava, cocoyam and vegetable plots were sole cropped

Table 4.1 Diversity of crops grown by the sample households in Ekwuloko and Edeke

	Ekwuloko HH				Edeke HH			
	M1	M2	M3	M4	E1	E2	E3	E4
White yam								
Water yam								
Cocoyam								
Cassava								
Sweet potato								
Root crops								
Maize								
Guinea corn								
Rice								
Cereals								
Cowpea					Riverine soil not suitable for			
Groundnut					leguminous crops			
Pigeon pea								
Leguminous crops								
Melon								
Okra								
Pepper								
Garden egg								
Vegetables								
Total number of crops grown by households	10	10	7	12	8	7	7	6

Shaded cell indicates that the household was cultivating the crop during the year

(only one crop on the same piece of land at the same time). Vegetables sole cropped were pepper and amaranthus (spinach or green leaf). There was no particular pattern observable among the four farmers in Edeke in relation to the proportion of land intercropped, except for the preponderance of intercropping in the late season relative to the early season. This can be explained by the presence of the white yam/water yam intercrop over the two seasons and the fact that there were fewer plots of cereals and vegetables in the late season because of flooding. The annual flooding in Edeke enables a virtual monoculture of yams and the crop can be grown continuously in the same site for as long as six years. This would not be possible in Ekwuloko. There

(a)

(b)

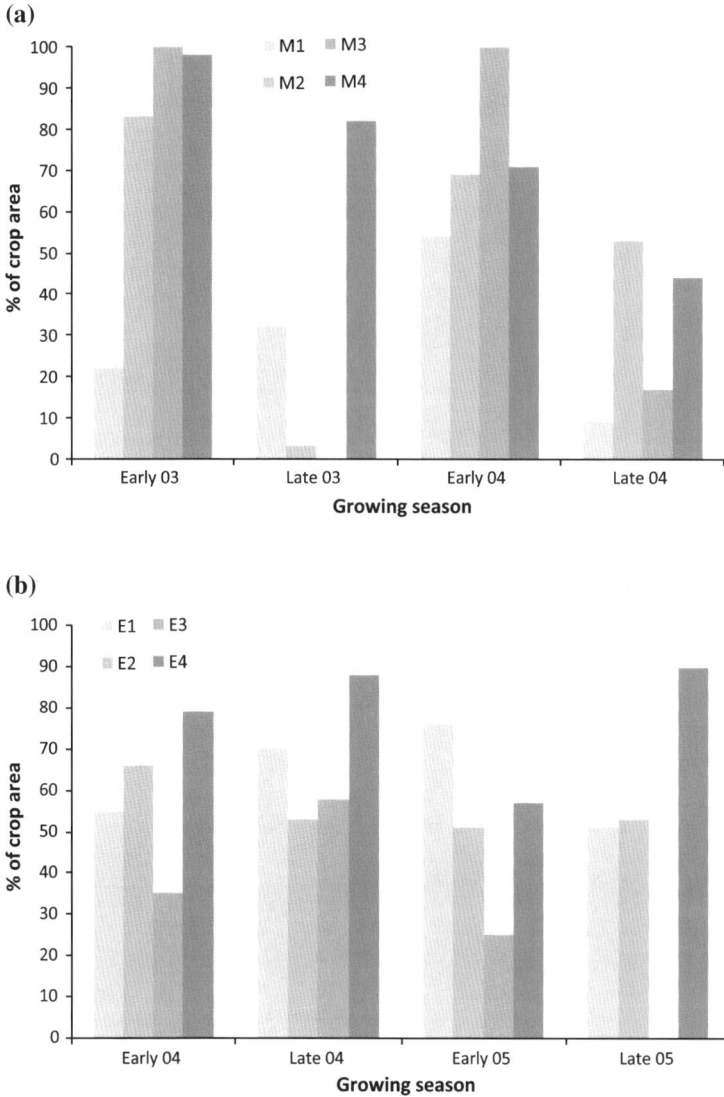

Fig. 4.9 Percentage of cropped area under the intercropping system. **a** Ekwuloko. **b** Edeke

was intense cultivation by the four HH in Edeke with more land planted in the early season than in the late season. Early season cultivation especially of yams, rice and vegetables usually begins in November; much earlier than is possible elsewhere in Igalaland. As the yam and water yam carry over to the second planting season there is proportionately more intercrop in the ground than in the first season.

As intercropping is common in Igalaland it is understandable that the land allocated to individual crops did not add up to 100 % (a plot size of 2,000 m^2 grown to maize and cassava will be recorded as that area for both crops—not half each). While such proportions are crude measures they can however be instructive. Indeed a comparison between Ekwuloko and Edeke as shown in Fig. 4.10 provides some interesting insights. The percentage of cultivated land will, of course, fluctuate as the area under cultivation changes, but it is a reasonable indicator of the importance placed on a crop by a farmer. Cropping systems in both villages are predominantly root crop-based which is reflected in the relatively high percentages for yam, cassava, sweet potato and cocoyam, but this is more so for Edeke than Ekwuloko. Maize is also important in Ekwuloko as indeed are the legumes.

Cassava-based systems marginally dominated over yams in the Ekwuloko sample while fopr the Edeke farms yams dominated (white yam, *Dioscorea rotundata*,

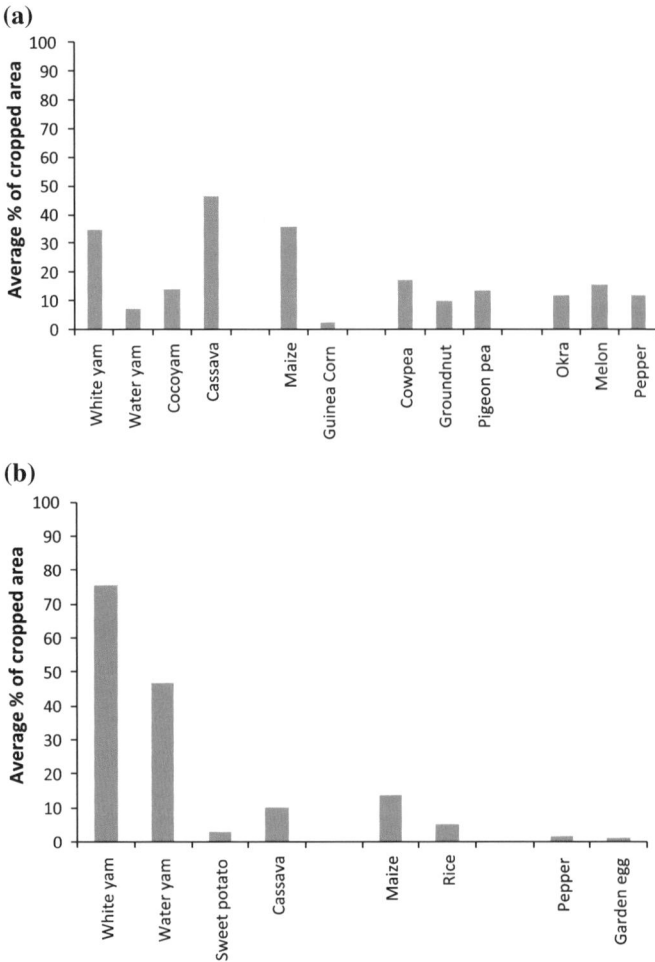

Fig. 4.10 Crops grown in Ekwuloko and Edeke. a Ekwuloko. b Edeke

and water yam, *D alata*). For all four Ekwuloko HHs white yam occupied a significant proportion of the cultivated land area, although the variation was from 13 to 60 % (typically higher in the late season as most cropping is concentrated in the early season). For the farmers in Edeke this percentage was much higher, with on average some 60 % (early season) and 90 % (late season) of the cultivated land being planted to white yam. Yam was clearly important to these farmers although it is not the only crop they coveted. The figures for cassava were much lower in Edeke than in Ekwuloko. These differences can be easily explained. Cassava requires much less labour than many other crops and is useful as a last crop in a rotation before returning land to fallow (cassava can do well on poor land). The production costs of cassava are also low compared with those of yam. However, cassava products regularly fetch a low price in local markets as well as being low in nutritional value, as they comprise little protein, but high in carbohydrate. Cassava is relatively new in Igalaland and hitherto yam was considered the staple. With declining soil conditions, the disappearance of many varieties of white yams as well as their relatively high production and maintenance costs, cassava grew in popularity in Igalaland. Its biggest problem is well known to be the relatively low and fluctuating market price. In Edeke where the land is good, the farmers have less motivation to resort to cassava. For them yam has better nutritional value and brings a far higher and relatively consistent market price.

Thus farmers in Ekwuloko do not only grow the crop with the highest economic value (such as yam) but also diversify their cropping to allow for their own consumption needs and to provide options to improve soil fertility. This enables them to manage their resources such as land and to provide some insurance against crop environmental and/or market failure. Overall this pattern is a classic indicator of poorer soil fertility in Ekwuloko relative to Edeke, with farmers having to diversify and grow crops which can cope with poor soil (cassava) and enhance it (legumes). But it has to be stressed that the cropping pattern identified for the four Ekwuloko HH farms along with the underlying rationale as explained by the farmers is common throughout much of Igalaland and was not surprising to DDS. Soil fertility management was not their only concern. Igbo HHs use legumes and vegetables extensively as food ingredients. Also it is noteworthy that substantial land ownership generally does not necessarily imply extensification of cropping (cultivation of large areas at low crop densities). Ironically, such extensive land ownership can go hand-in-hand with indicators of intensification (cultivation of small areas at higher crop densities— usually with additions of inputs) as farmers try to get the most out of plots closer to the village thereby avoiding long-distance travelling to their farms. Closeness of plots and farmland to compounds is now critical to ensuring the crop is not stolen. Food security and human security are thus becoming inextricably entwined.

A further interesting question is whether there was any difference in cropping patterns between the Igala and Igbo HH in Ekwuloko? There were similarities of course; the percentage areas for crops such as yam, cassava, pepper and cowpea were similar for the Igala and Igbo HHs. But there are also differences; percentage areas of maize, groundnut and melon were higher for the two Igbo HHs, while areas of cocoyam and okra were much higher for the Igala HH. The difficulty is one of reading too much into these differences given that crop areas of only two HH from each of the ethnic groups were assessed. Differences will reflect HH dietary taste as well as culture and the type of land to which HHs have access. Cocoyam, for example, requires good quality land, and Igalas generally have a major preference for okra in their diet perhaps more than Igbos do.

Thus overall it has to be said that the information collected on land ownership and cropping patterns presented no real surprises to DDS and confirmed patterns that they had seen elsewhere. The biggest surprise was the scale of land owned by the two Igala HH in Ekwuloko, but given that those selected for the SLA were 'senior' then in retrospect one would expect this.

4.5 Natural Capital: Trees

Trees are an important natural asset in Igalaland for a number of reasons. They can be an important source of income, particularly for women, they help maintain biodiversity by providing a range of habitats, give much welcome shade and protect the soil. Some of these advantages are directly linked to livelihood whereas others, such as maintenance of biodiversity, are indirect. Indeed the main cash crop of Igala is arguably palm oil from the oil palm tree (*Elaeis guineensis*) rather than field crops. Oil palm has its centre of domestication in the Niger/Benue flood plains, and once slavery was abolished the oil palm in Nigeria became a major export of interest to British companies (*Lever for example*). Oil palms provide the raw material and by-products for a range of local industries suitable for women, and like yam it is a significant cash crop in the local economy. Some of the

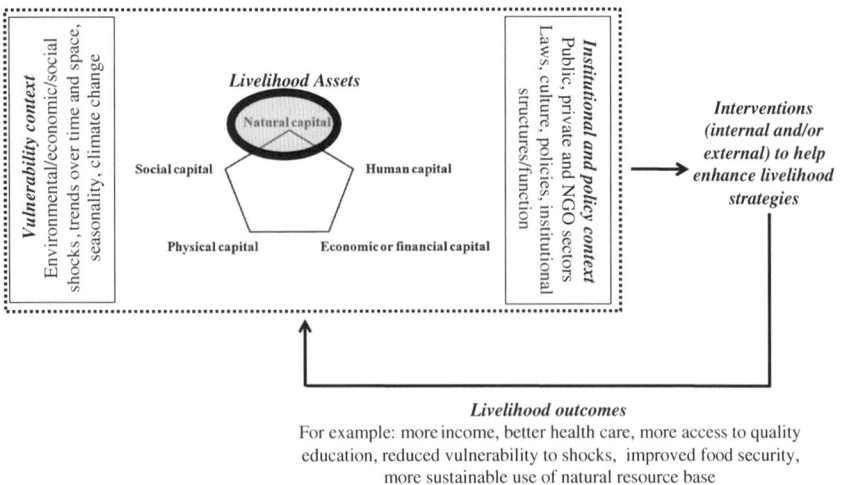

Fig. 4.11 Natural capital (tree crops)

produce is also consumed locally especially in the local food industry. Indeed a much sought after delicacy is boiled yam eaten with a palm oil sauce.

Traditionally, tree crops can be a contentious resource as they relate to land ownership—ownership of trees can imply ownership of the land upon which they grow. Economic trees are a long-term investment and most tree crops can take years to reach full maturity (even the fast-maturing oil palm varieties can take up to five years). However, once established they require little maintenance, produce can be lucrative and in most cases they have a high nutritional value. Women in particular are interested as they do the marketing of tree crop produce; they use oranges and palm oil to improve nutrition at household level. These resources provide back-up income for school fees and other essential household needs (medical and clothing) where income from arable crops has proved less reliable. Prices for many crops, especially cassava, fluctuate depending on supply and demand.

Other studies of tree crops in Igalaland facilitated by DDS (Morse et al. 2000) have shown that fathers were planting tree crops especially oranges, bush mango and oil palms for their daughters and sisters as part of their dowry. Women had access to such trees long after their fathers had passed away even though women were not able to own land. However access to the produce of such trees extended only to the lifetime of that woman. In trying to make provision for their daughters and sisters men saw that trees were a reliable source of income and their initiatives have given rise to cultural transformations within many villages.

A summary of tree crop ownership and revenues for the HH in 2004 (Ekwuloko) and 2005 (Edeke) is provided as Table 4.2. These figures refer to the fruit trees (orange, guava etc.) and also oil palm, the two types of locust bean and trees such as gmelina and teak which are a source of timber. There are a number of interesting points that emerge from this table. Firstly, the Ekwuloko HHs had far more trees than Edeke HHs. In Edeke the average ownership of tree crop stands was 60 while in Ekwuloko it was 1,065. The difference was largely accounted for by oil palm; the Ekwuloko farmers often had many stands while the Edeke farmers had none. The regular flooding of riverine soils does not provide a suitable environment for oil palm.

Table 4.2 Tree crop ownership and revenue for Ekwuloko and Edeke

	Number of stands (04)	Revenue (04)	Other trees also economically important in the local context
M1	1,148	55,000	Oil palm, locust bean, gmelina, teak
M2	79	39.700	
M3	1,099	None	
M4	1,935	18,500	
	Number of stands (05)	Revenue (05)	
E1	24	None	Bamboo
E2	37	35,200	
E3	49	24,000	
E4	130	17,500	

Trees contributing to revenue in both villages include orange, guava, mango, cashew, banana and pawpaw

However, it can be difficult for farmers to estimate the exact number of stands of some species (locust beans) as they are self-generating. Hence figures for Ekwuloko in particular could well be underestimated. The good news in terms of sustainable livelihoods is that all trees mentioned in Ekwuloko and Edeke are inexpensive to produce as well as fast growing and thus provide a viable alternative source of revenue for the HH.

As expected in Ekwuloko, the two Igala HHs (M1 and M4) had the largest holding of tree crops compared to the two Igbo HHs (M2 and M3). This links immediately to the land ownership issue already discussed. People who rent land are often loath to plant tree crops, as these automatically become the property of the landlord. Yet surprisingly all four farmers referred to themselves as 'owners' of the tree crops. This was unusual, and that was the reason why people distinguish between the 'owner' of trees and 'controller'. A controller was someone who had access to and control over produce and received the full benefit of planting the trees for him/herself even though they did not necessarily own the trees. Hence a husband would 'own' the trees and the wife would 'control' them. In Ekwuloko the Igbos planted trees for themselves and controlled their management and produce but if they left, ownership reverted to the landowner. No relative could come back to claim them later. In Edeke tree crop ownership was more uniform across the four households.

While much of the produce from tree crops was consumed by the HHs, it has to be noted that income from trees was significant for some of the households (up to N55,000 in the year). Another surprise was the local nature of the markets. Ekwuloko respondents reported selling their produce through their local market and this was unexpected to DDS given that the market did not look all that big (markets in Igalaland—like much of Nigeria—operate on a four day cycle). It was anticipated that more of the produce would be sold in the larger markets of Odolu or perhaps even Idah and Nsukka. However, upon further investigation it was clear that Ekwuloko market is deceptively large. Indeed, it was the only major market for some distance and Igala and Igbo traders travelled there to purchase agricultural produce and manufactured goods such as pestles and mortars. The Edeke HH had the advantage of proximity to Idah and that market, one of the largest in Igalaland, unsurprisingly featured heavily in their responses.

The complete absence of kola nut from the tree crop culture was a further surprise, especially in Ekwuloko. The Edeke soil is not conducive to its growth but the environment in Ekwuloko certainly is. Kola nut is an important tree crop from a traditional perspective, given as a sign of welcome at traditional feasts and meetings, and it tends to fetch a good price. For a village where it was expected to be in abundance its absence was a mystery. When questioned the farmers referred to a lack of 'know how' in growing it. Given the Igbo presence in Ekwuloko it was also surprising to discover the low numbers of banana/plantain stands. Bananas and plantain are important as they form part of the staple diet, and local growing conditions seemed suitable. The main reason proffered for not producing it was that land was rented. Other oddities included the absence of bush mango and low numbers of locust beans. Even the most soil-depleted places in other areas where DDS had worked had more of these tree crops than did Ekwuloko. A rather aggressive bush clearance must have taken place at some time, and the repercussions must have lingered to the time of the SLA. There is a mere semblance of teak and gmelina and fuel wood will surely be a problem for the future as the population grows. For a more balanced future economy, fast growing fuel trees needed to be encouraged as they were a source of income contributing to the overall well being of the economy and the environment. The HHH M4 already indicated he was taking up this challenge. However, it has to be said that while there were gaps in local knowledge of tree crop planting and maintenance, the composition of the community in Ekwuloko did not lend itself to tree planting. Fixity of tenure did not exist for a significant proportion of the Ekwuloko population—the Igbos. The results showed HHH M1 as benefiting well from his tree crops. The Igbo household (HH M2) grew fruits with a view to improving diet and subsequently sold the balance. HHH M3 (Igbo) was not as committed to such matters as he had less land and the family bi-locates between houses in Ekwuloko and Nsukka.

Tree crops grown in Edeke were related to the economy of hunting and fishing and particularly the latter. A limited number of trees were grown in the uplands, which the lowlands could not support, and it should be noted that all four HHs in the SLA rented their land. Oil palm was the most popular in the more elevated parts of Edeke though most of these were self-seeded. Fruit trees especially the improved varieties were now to be found especially oranges, guava and mango. Cashew was also to be seen but grew in a vegetatively condition, a sure sign of high quality land. Cooking banana and new improved plantain varieties were in evidence too, with demand for planting material exceeding supply. Good varieties of pawpaw were to be found in almost every compound, a fruit that is particularly useful in traditional medicines now making their return in Edeke. Such crops were also being promoted by DDS as part of nutritional and environmental programmes. Edeke has long since benefited from programmes promoting tree crop maintenance and even if the number of trees growing there was limited, good quality fruit obtained in households.

Calabash trees were popular in Edeke due to their role in the local fishing industry. Bamboo was to be found everywhere, considered vital to the lives of fishermen and women. It served as useful material for kitchens and yam barn, *atakpas* (meeting places) and houses.

Overall the results with regard to tree crops provided DDS with both the familiar as well as a number of surprises. The relative paucity of some tree crops in Ekwuloko—as evidenced by both the returns for the four HH and general observation within the village—was the most surprising result for a village in Igalaland.

4.6   Social Capital: Networks

Social networks are an important aspect of Igala and Igbo society. These take many forms, ranging from faith-based groups, youth groups, labour rotational groups, and savings/credit groups to entertainment clubs. They were therefore important instruments, worthy of consideration in the SLA. A significant social unit in Igalaland is the clan (an extended family unit), and clans tend to have a

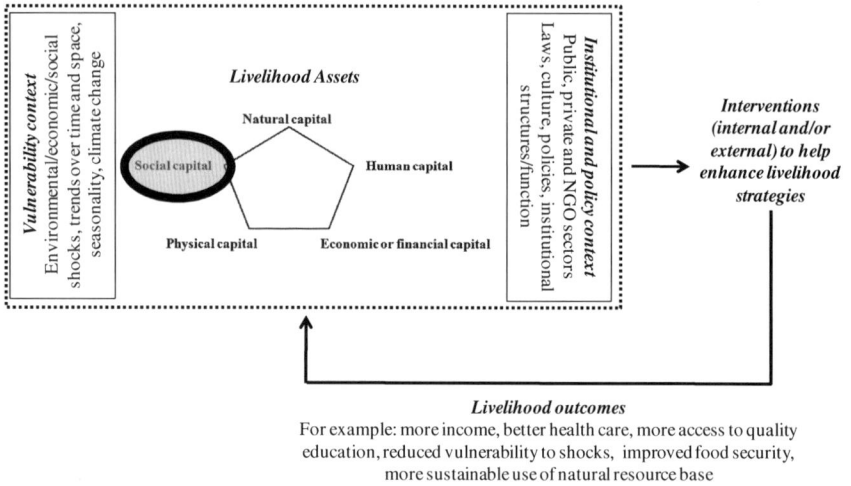

Fig. 4.12   Social capital

clear spatial pattern in the layout of HH buildings and meeting places. Usually there is a central meeting place for each clan called the *atakpa* (can be loosely translated as 'parlour' or 'meeting place' in English). However, in this respect, Ekwuloko turned out to be unusual when compared with other village studies conducted by DDS. The Ekwuloko community can trace their ancestries back to clans in Idah of which they are still part, and indeed Ekwuloko has its clan network linked to the Royal Clans. Both ethnic groupings in Ekwuloko have their own clan meetings and as need arose, all met with the *gago* (chief) for the resolution of problems or decision making. Indeed the large diversity of societal membership amongst the HH samples in Ekwuloko and Edeke was striking.

A summary of societal membership in Ekwuloko and Edeke is provided as Fig. 4.13 where groups are pooled under broad headings. The three headings chosen here are faith-based and secular (non-faith based) groups, local to the village and non-local (national for example) and agricultural and non-agricultural. The only HH in both villages with a relative paucity of membership in groups was M3, but the situation here may be because his wives and older children did not for the most part live there. The other Igbo HH of M2 appeared to be much more embedded in the local, although still with dominant membership in non-local groups. The two ethnic groups in Ekwuloko also had their own meetings and in instances where decisions needed to be made at village level the gago summoned the head of all groups. For the two Igala HHs in Ekwuloko, membership of local groups tended to dominate over non-local, while the opposite was true for the Igbo HH of M2. Indeed the Igbo HHs were more inclined to membership of faith-based groups.

There were some interesting differences between Ekwuloko and Edeke group membership highlighted in Fig. 4.13 and this was particularly noticeable with regard to membership of faith-based groups and to a lesser extent with the

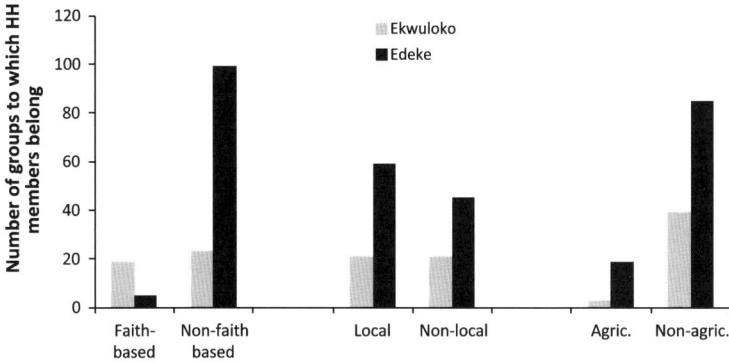

Fig. 4.13 Summary of group membership in Ekwuloko and Edeke. Figure shows membership in three different types of group: faith-based, local and agricultural

agricultural and non-agricultural categories. Membership of faith-based groups was more dominant in Ekwuloko than in Edeke; three of the Edeke HH were Muslim while all Ekwuloko HHs were Christian. Faith-based groups tend to be more numerous within Christian Churches than in Islam.

The Catholic Women's Organisation (CWO) is an active organisation of Catholic women with branches in every diocese and parish throughout the country and linked with its sister organisations in Europe and the USA. There are exchange visits between parishes in Idah diocese and Fulda diocese and parishes in Germany. The main purpose of the CWO is to help its members be more Christian and Catholic through learning about, and putting into practice the principles of the Christian and Catholic faith.

The Catholic Youth Organisation of Nigeria (CYON) was also to be found in Ekwuloko as it exists in every parish and diocese throughout the Federation of Nigeria. It too has international links but is usually most active in secondary schools and sometimes in tertiary institutions. Membership of such organisations is typically seen as a means of helping with character formation instilled mainly through life coaching skills which would help youth help themselves and others for the rest of their lives. They became aware in theory at least that on—going formation is essential if one was to live a Christian life to the full through being our brothers' and sisters' keepers.

Besides the CWO and CYON there were six other active Catholic organisations in the Ekwuloko community. These were the Knights of St. John, the Knights of St. James, the Legion of Mary, St Anthony De Padua Society, St. Augustine Society and St. Teresa's Society. The Knights of St. James and the Legion of Mary were part of the Catholic International network. The others were to be found in practically every parish and diocese which is likely to have more of such groups than found in Ekwuloko. The paucity of such societies would indicate that Catholics were fewer in number than Protestants here. Society members helped in Church maintenance and had different duties depending on what activities were happening. There was no Catholic Priest in the village (Ekwuloko was part of

Akpanya Parish) but there was a resident catechist. At the time of the SLA one of MM2's sons was a seminarian. The Protestant groups in Ekwuloko were extensions of national evangelical churches, some with an international link. Members met each Sunday for service and other activities such as Bible studies. They did not seem to engage in any specific social activities.

Religion forms an important part of life in every HH in Igalaland and much of the social life of the Igala communities revolves around meetings based on the religion to which they belong. Members of the same family may belong to different religious denominations and rather than being divisive this is often a unifying factor within communities. Igbos tend to be Christian in the main while Igalas can be either Muslim, Christian or traditional believers. This fact ensures that what is learned in one group is also shared with other members of the same family and community. The diversity works in many ways as the following example relayed to the authors by DDS staff shows. In the course of DDS implementing a rain harvesting programme in an Igala area, there was a request for three rain harvesters from the same village. On investigation it was discovered that one was for the Muslim community, another for the Protestants and a third for the Catholics. The population warranted three as there was almost a mile between each project proposed. The three local contributions (money paid by the communities to DDS) were complete and the three leaders (one from each denomination) approached DDS to explain the reason for three applications. Initially when the community decided to engage with the construction of a rain harvester there was dissension as to where the actual project should be located; would it be in the centre of a rather large village or if it were to be in a church or mosque compound then which one would win out in the end? There was stalemate. It was at this point the three leaders intervened to resolve the issue. The population alone warranted three projects as this village had a population in excess of 5,000 people. Water was by far the greatest need so many sacrifices were made by men, women and children to reach the required local contribution. The most efficient way of collecting this was through the religious leaders who were united in their efforts to improve conditions for all members of the community. This collection generated some competition and the target was reached in record time and the three projects were successfully concluded. This is just one of many examples in the experience of DDS where religious groups are a vital source of social capital which not only gives meaning to their lives but also helps them engage more meaningfully with each other and the wider society. However, there are also occasions when dissension has had the opposite effect to what happened here.

Infrastructure projects such as water provision outlined in the previous paragraph are by no means the only means by which DDS engages with its members via faith-based groups. From the perspective of DDS the religious aspects of group membership, which impart important values to its members, are now beginning to ensure that members are part of discussions and debates that encourages them to think for themselves and not be mere passive recipients of information or pawns in bringing about undesirable change. Indeed in Igalaland religious and cultural values are often similar and help members to show respect for their own culture as well as what needs to be challenged. In recent times most groups have benefited from information on critical issues and foremost among these is HIV/AIDS. Disbelief at first, has given way to a realisation that the pandemic is for real and that a change of behaviour is called for where needed. HIV/AIDS is by no means the only issue raised by DDS via such groups. It has also been possible to educate women and men of the evils of trafficking young girls and women through making them aware of the different guises such 'modern-day slavers' employ and thus helping to prevent this from happening. They also learn about what can be done in the case of people whom they believe may have been trafficked. Women become more aware of their rights and how to protect these rights. Men are helped to be more in tune with such issues and this information and the discussion it provokes is gradually leading to profound changes in relation to women's rights. Widows are treated with more respect and such mourning customs that demand women to sit on the floor in her room alone have been replaced by more consoling and comforting rituals. Other issues such as 'Care for the Earth' have been raised by DDS via religious groups throughout Igalaland and they see it as an efficient and sustainable means of education and awareness building. Women in particular welcome and cherish such input and not only respond but also invite people to give input. DDS has also known for long that such groups are an important means of recording change. What happens within them helps those who give and those who receive and all parties are gaining as everyone has an opportunity to be both a donor and a beneficiary. Progress is incremental but enduring.

Many of the non-faith based organisations found in the two villages have branches in Idah and throughout Kogi State. *Ufedo kpai Udama* is an example of such an organisation. There were approximately seven women's organisations mentioned by the HH in Ekwuloko, and were for the most part groupings of women who came from villages close by and who formed a solidarity group for their own well-being. Such women's groups are common and are to be found in every town and village in the country as a whole

Youth clubs and associations were mentioned by a number of HH members. Youth clubs assist the villagers in what is known locally as 'credit labour'. Any farmer in need of hired labour but short of capital could engage a youth group. At harvest the members are paid in cash at a price agreed upon at the time of hiring and usually reserved for Christmas celebrations. These same groups provide communal labour for land clearance or harvesting free of monetary charges but are compensated by the community with food and drinks at Christmas. At the time of the SLA there were two such groups in Ekwuloko; one named the Christian Youth Group the other Ekwuloko Youth Club. Age of membership for these groups ranged between 15 and 40 years. However there was a preponderance of members in the upper age limits, a common phenomenon throughout this region.

Only two of the Ekwuloko HHs (M2 and M4) mentioned membership of a farming association as important whereas all the Edeke HHs included in the SLA had members of farming associations. It could be argued that membership of social groups such as farmer's associations might have been an indicator of interest in farming as well as a willingness to try new ideas. However, farmers' associations are to be found everywhere in Nigeria and in practice seem to have little connection with improving agriculture as they were often political.

Only one of the Ekwuloko HHs (M3) and two of the Edeke HHs (E3 and E4) mentioned any association with a political party (PDP, the ruling party in Nigeria at the time of writing) as important. It was not surprising that a 'Drivers' Union' was mentioned in the lists as HHH M4 spent many years as a driver, an opportunity rich with possibilities that later helped him develop his entrepreneurial skills.

Figure 4.14 shows the mean number of societies to which HH members belonged. Interestingly the mean number of social groups to which members of the four Edeke HHs belonged varied between 3 and 6, and these figures were higher than for Ekwuloko. In Ekwuloko the number of social groups to which an individual from any of the 4 HHs belonged varied between none and 5. A number of factors could have determined this. Age is very much wrapped up with status in Igala and Igbo societies and thus the household head and his wife (wives) would have had status to maintain, and membership of social groups might have been one vehicle for achieving this. Such status might be of lesser importance for sons and daughters and perhaps even less again for friends who were only temporarily residing in the village.

Generally the diversity of social networks in Ekwuloko and Edeke was encouraging and no doubt an asset in helping to make livelihoods sustainable. They provided an immediate and efficient means of communication where there was informal discussion with problems and possibilities highlighted. While societal membership is well known throughout Igalaland, and the societies that emerged from the SLAs were not especially surprising, the differences between the HH of the two villages were not expected by DDS.

4.7 Physical Capital: Assets for Income Generation

Access to land for agriculture is obviously important within a largely agrarian society such as those represented by Ekwuloko and Edeke. Hence it was reasonable for DDS to spend time in understanding the land tenure and agricultural

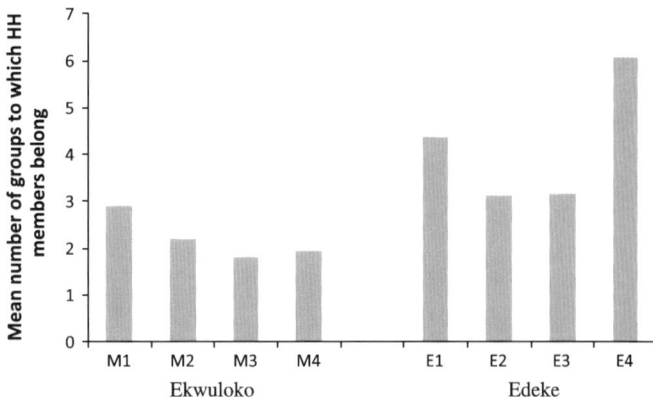

Fig. 4.14 The mean number of groups to which HH members in Ekwuloko and Edeke belong

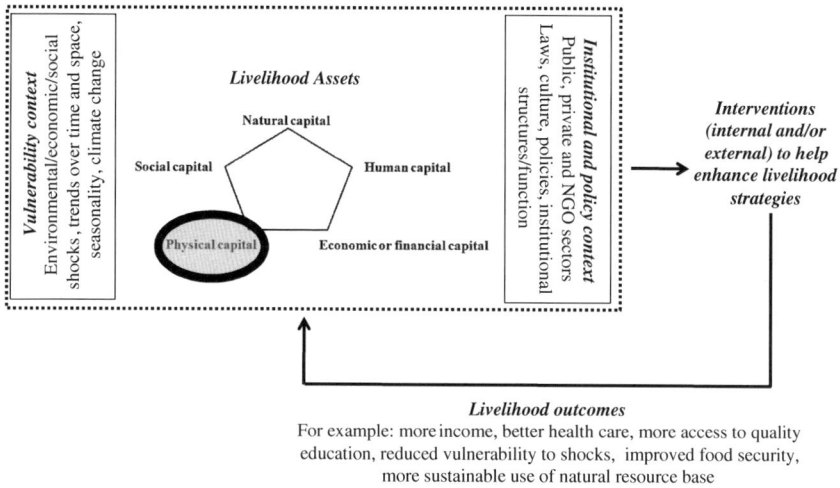

Fig. 4.15 Physical capital associated with income generation

systems. The soil types and local environmental conditions were quite different in the two villages, hence it was not surprising to see differences in terms of crops, cropping systems and trees. However HHs have access to other ways of making an income besides agriculture. Indeed all the HHs included in the SLA talked about income generation outside of agriculture as being important and were willing to provide examples such as trading and income from rental of land and machinery. But judging how important these are can difficult as such information generally tends to be sensitive. This point has already been made in Chap. 2 and is a familiar issue to DDS. As a result it was necessary to triangulate all information with observation and one reliable method of addressing this within the SLA context was to catalogue physical assets (tools, vehicles, machinery, livestock, buildings etc.) available to the HH and assume that this was related to their relative wealth and the options they may have available for income. This is not a perfect solution, of course, but it does provide some clues.

A summary of the total value of the HH assets for the samples in the two villages is provided as Fig. 4.16. The most striking point to emerge was the low value of assets for M1 compared with the other HHs in Ekwuloko and Edeke. With a total asset value of less than N200,000 this HH would appear to have had only a fraction of the wealth of the other three HHs. It should be noted, however, that M1 had extensive areas of land that he rents, and thus could provide a source of income. At the time of the study he had 15 tenants who paid him Naira 1,500 rent each per annum. These tenants also did compulsory labour for him each year. Secondly, the value of the Edeke HH assets was far below those of Ekwuloko HHs with the one exception of HH M1. Much of this difference was due to HH M2, M3 and M4 owning many buildings within and outside Ekwuloko. By way of contrast the Edeke HHs were more 'fixed' in Edeke, with few (if any) assets outside the

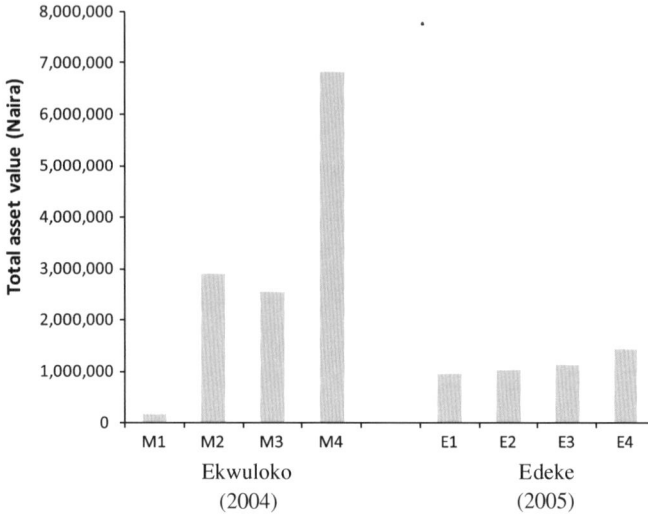

Fig. 4.16 Total physical assets (not including value of land) owned by the four HH in Ekwuloko and Edeke

village. It was worthy of note that the variation between the four Edeke HHs in terms of their asset value was much less than that of the Ekwuloko villages. The difference between E1 and E4 was some N800,000 while the difference between M1 and M4 was more than N6.5 million. Even between M4 and M2 (ranked 1st and 2nd respectively in asset value) the difference was N4 million, and M4 is by no means the richest HH in Ekwuloko.

The wealthiest HH of the eight was M4, whose combined assets were worth more than the total of the other three HHs in Ekwuloko put together and the same would be true of the four Edeke HH. One visit to his house was sufficient to confirm this. The HHH of M4 was an entrepreneur *par excellence.* He was as well a perfectionist as evidenced in his farming practices. He was committed to this business so it was easy to see how he could accumulate wealth; he was painstaking and thorough in all he did and not one to miss out on any opportunity. However, it should be noted that the high value of M4 assets was in part due to the HHH owning a house in Idah. He traced his ancestry to one of the clans in Idah so he made the most of that opportunity by having a house there. Houses in Idah then had a high monetary value. The Idah house was valued at N3,000,000; therefore nearly half of M4's assets were tied up in the Idah house. Nevertheless, even without this asset the valuation of M4's assets was more than any of the other three HHs. The differences can be attributed to differing valuations of houses between Idah and Ekwuloko. The term 'house' was a loose one and would typically cover a number of buildings not just one. While individuals would naturally tend to perhaps over-value their own buildings the DDS researchers confirmed valuations via a neutral source.

A second complication with asset valuation was linked to location. This point has already been noted for HH M4 with regard to the ownership of the house in

Idah. The same applied to a lesser extent to the two Igbo households in Ekwuloko, but especially to M3. The house valuation of nearly M2,000,000 covered two properties: one in Ekwuloko and one in Nsukka. Also, the valuation of N150,000 for two grinding mills covered the value of one at Ekwuloko and one at Nsukka. This further highlighted the 'bi-locational' nature of three HHs included in the study. The only one with a sole presence in Ekwuloko was M1.

It was another matter when houses and buildings were taken out of the equation and comparisons were made in terms of what could be referred to as 'productive' capital although this was not an easy distinction in the Igala context where many assets are flexible in the sense that they can be used to generate income, even if only in kind. Also, valuation of an asset may not be directly linked to its potential for income generation. Even so it can be useful and here it was taken in a narrow meaning of agricultural equipment, livestock/fishing, transport (necessary for marketing, paid employment etc.) and processing. The valuation of 'productive' assets under five categories is provided as Fig. 4.17. The greater comparative wealth of some of the four Ekwuloko HHs (notably M2 and M4) compared to Edeke was again evident in the 'productive' assets. Indeed the productive assets of M4 were almost equivalent in value to the combined productive assets of all four Edeke HH. However, Edeke HHs had more in the way of agricultural assets, and this reflected the larger land areas they farmed and the dominance of agricultural products in HH income, a point that will be returned to later. In Ekwuloko the valuation of agricultural tools was much the same across the four HHs, with the high value of N152,100 for M4 being explained by the inclusion of a motor which showed M4 clearly had much less livestock than the other three HHs, probably further confirming the part-time nature of his farming activities. However, both M3 and M4 had a great deal of investment in transport, confirming their greater mobility. These two HHs also had a strong investment in crop processing.

How does this pattern of asset ownership seen for the sample of HH compare with the other HHs in the villages? It was neither possible nor necessarily desirable to carry out the same detailed assessment as presented here for the whole village, but it was possible to use some indicators of wealth across HHs in Ekwuloko to allow a comparison with the four HHs. This was not logistically possible in Edeke. A summary of the wealth ranking exercise for Ekwuloko is presented as Fig. 4.18. The three indicators of wealth employed were:

1. Total number of buildings owned by the HH (dwellings, kitchen, toilet, goat houses, processing buildings etc.)
2. Car ownership
3. Ownership of motorcycles and grindings mills (combined)

These indicators were in order of importance. Both the choice of indicators and their relative ranking were as determined by people in Ekwuloko and not by DDS. Figure 4.18 presents the ranking of the 111 HHs in Ekwuloko based upon these indicators. The range from top to bottom, particularly with the first indicator (number of buildings occupied by the HH) is substantial. Here it ranged from one to 10 buildings. However, it was obvious that many Igbo HHs in particular rent the buildings (albeit

(a)

(b)

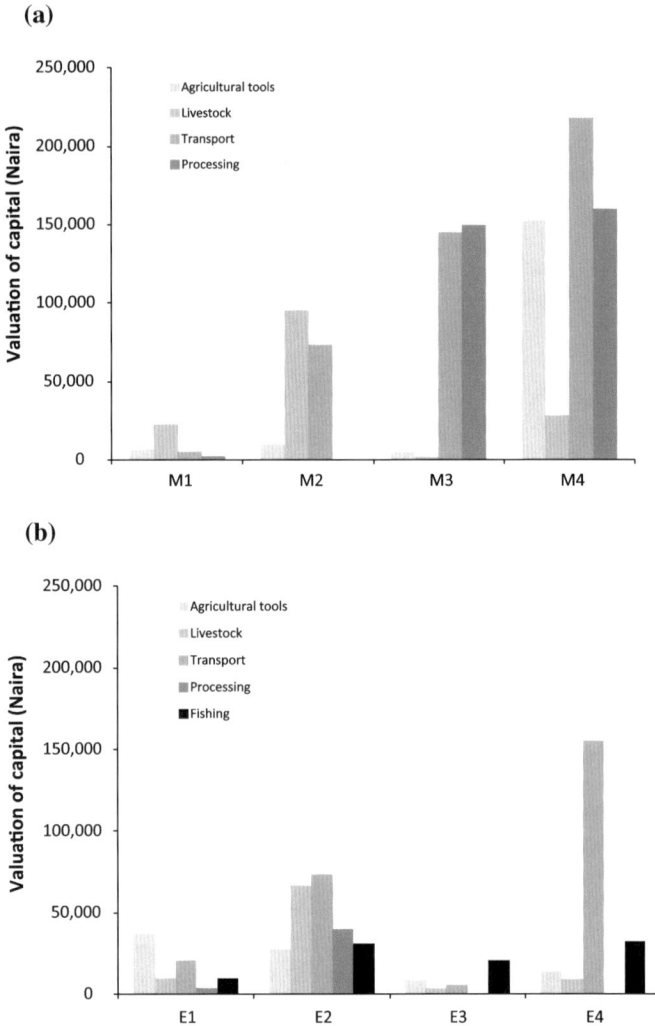

Fig. 4.17 Valuation of 'productive' assets within the categories of agricultural tools (cutlass, hoe, saw), livestock (hens, goats, sheep), transport (bicycle, motor bike, wheel barrow, canoe), processing (oil palm trough, grinding mill) and fishing (nets, hooks, spears). **a** Ekwuloko (2005). **b** Edeke (2004)

on a long term basis) rather than own them. Wealth ranking based on occupation of buildings either through ownership or rent can therefore be considered as somewhat crude. However, it would have been difficult and time-consuming to pursue ownership in greater detail. After consultation with key informants it was apparent that the results would have presented a rather biased distribution of wealth in Ekwuloko that favoured Igala HHs, markedly underestimating the wealth of Igbo HHs.

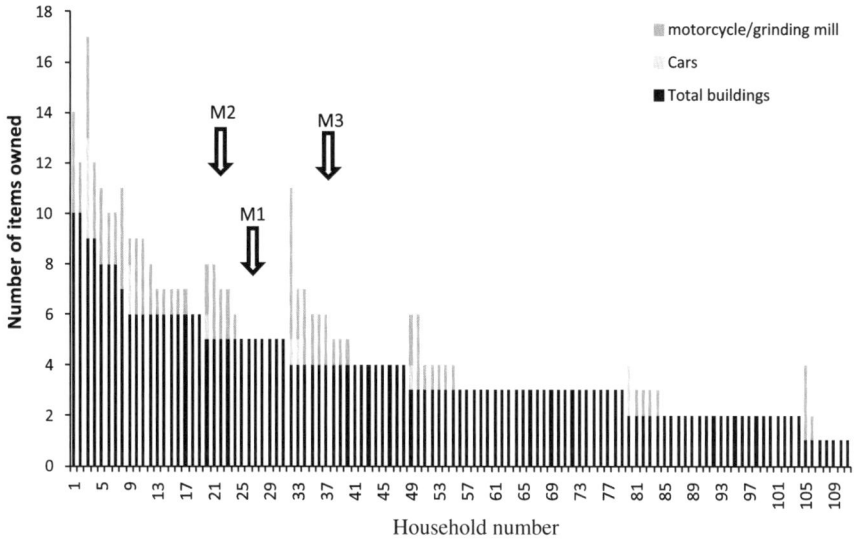

Fig. 4.18 Wealth ranking, based upon three indicators, of HH in Ekwuloko. Also shown are the relative placement of three of the sample HH within the village based upon these indicators. Wealthiest HH are to the left hand side of the graph

However, the four HHs selected for the research in Ekwuloko did not span all wealth categories in Fig. 4.18. One HH (M2) appeared in the wealthiest 20 % while two others (M1 and M3) appeared in the next category. M4 is not living in Ekwuloko and therefore does not appear in Fig. 4.18, but based on the same indicators M4 would also appear in the top 20 %. Hence all four HHs were in the wealthiest 40 % of the Ekwuloko HH population. As already mentioned, this was largely a reflection of the self-selection by the farmers in the village of four HHHs who would command respect and provide leadership. Therefore the fact that they were in the top part of Fig. 4.18 was not a surprise.

The figures in Fig. 4.18 can also be broken down in terms of Igala and Igbo HHs and the results are shown in Fig. 4.19. Of all the HHs in Ekwuloko, Igala HHs tended to be wealthier than the Igbo, based on the three indicators. This was not surprising given that Igalas own the land and had more potential to develop their assets and build up their wealth. However, some Igbo HHs were not so far behind; the third wealthiest HHH in the village was Igbo.

It is also important to note that there was no difference in distribution between male and female headed HHs in terms of the wealth indicators. Females headed some 14 % of HHs in Ekwuloko, and as mentioned previously the trend in Igalaland was towards having more female HHH as males migrated looking for work. Unlike ethnicity, there was no apparent association between wealth and gender of HHH. Female-headed HHs were as likely to be classified as wealthy or poor as male headed HHs. The explanation for this equality is relatively straightforward. In Igala and Igbo culture a female-headed HH can arise when the husband is

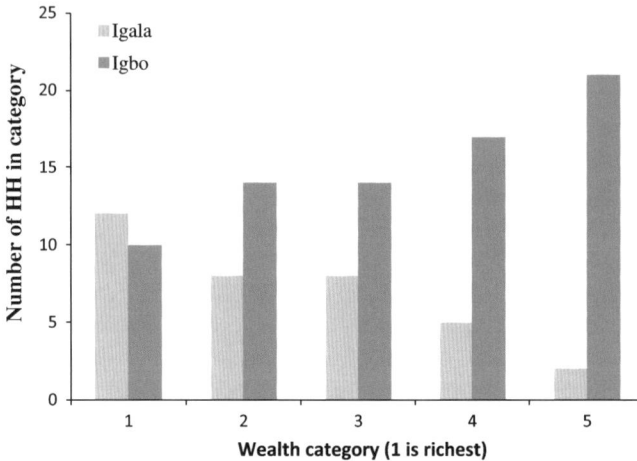

Fig. 4.19 Distribution of Igala and Igbo HHs (based on ethnicity of HHHs) amongst 5 wealth categories

working away from the village, perhaps in paid employment elsewhere or when a husband dies. If the husband dies then assets, including land, are passed to the eldest son if he is old enough, who, strictly speaking is the new HHH. However, if the eldest son is too young to assume this responsibility, the family of the late husband is entitled to take all assets but is obliged to look after the widow and her family. Nowadays this law is being relaxed as education increases and as there is no guarantee that the deceased husband's family will take care of the widow and her family. The 'acting' HHH is then the eldest female, typically the wife. If the husband is living away he is still technically the HHH but the wife acts in that capacity. In either case it is unlikely that the wealth indicators will detect any difference.

4.8 Financial Capital: Household Budgets

As well as an examination of assets as a means of assessing wealth in the community, DDS has in the past found it useful to explore income and expenditure by a HH. As difficult as the cataloguing of physical assets may be, it is nothing compared with attempting to assess income and expenditure as HH members are understandably loath to reveal such information. However, in general they are more than willing to set out their expenditures as a way of stressing their 'suffering'. The result is typically a much higher value for HH expenditure than for income as all details pertaining to health costs, school fees, food etc. can be remembered. Attempts to exaggerate costs is common but can be readily checked when compared with other areas by both DDS Igala and Igbo staff and key informants. Expenditure can be a reasonable proxy indicator for income if it is assumed that a

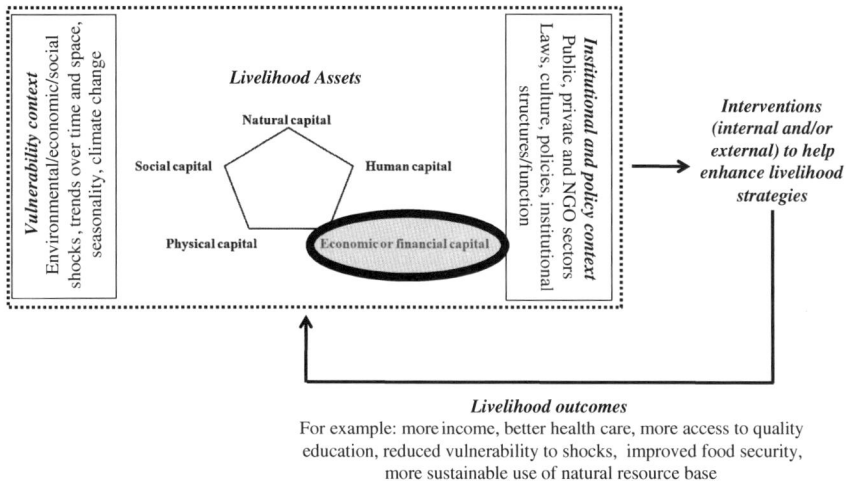

Fig. 4.20 Financial capital

HH account is more or less in balance, and this can be checked against asset ownership to see if there are signs of investment. However, the value of HH income/expenditure analysis rests more with the information it provides as to the main sources of revenue sinks and their relative rank rather than the absolute amounts.

Table 4.3 is a summary of an income: expenditure budget for the four HHs in Ekwuloko and Edeke. The figures are presented as absolute amounts of income and expenditure as well as a percentage of the total. Also shown are the balances (income—expenditure). The budgets raise a number of interesting points. The results from Edeke were either negative or only marginally positive. This contrasted with the more favourable results for Ekwuloko (except for the special case of M2 where there was a relatively high expenditure on health care for that year). Of course these were only the declared amounts and it is likely that in both villages there was an under-estimation of income and an over-estimation of expenditure. Thus the balances were likely to be higher than actually declared. Even so, the negative balances suggested that the four Edeke HHs were relying on credit to maintain their production and indeed conversations with the HH and others in the village in a similar position suggested that was the case. All stressed the importance of credit for farming and fishing far more often than did respondents in Ekwuloko. During the interviews the four Edeke HHs claimed to have borrowed the following amounts for farming for the 2005 growing season:

E1 N300,000
E2 N300,000
E3 N250,000
E4 N40,000

These figures were broadly in tune with the negative balances for the HHs, although it should be noted that interest payments on loans were high—often as

Table 4.3 HH budgets (income, expenditure and balance) for Edeke and Ekwuloko

(a) Ekwuloko (2004)

Income

	M1	%	M2	%	M3	%	M4	%
Rent	74,700	27	7,200	1	0	0	437,700	39
Crop sales	91,300	33	371,700	50	89,150	33	395,500	35
Livestock sales	15,400	6	0	0	0	0	200,000	18
Tree crop sales	55,000	20	39,700	5	0	0	18,500	2
Paid employment	41,500	15	324,000	44	180,000	67	68,000	6
Others	1,500	1	0	0	0	0	7,700	1
Total	279,400		742,600		269,150		1,127,400	

Expenditure

	M1	%	M2	%	M3	%	M4	%
Food	47,000	16	203,400	11	47,220	22	181,400	18
Religious fees	20,000	7	76,750	4	28,820	13	35,600	3
Transport/building/maintenance	41,200	14	458,000	25	85,820	40	192,400	19
Community fees	11,500	4	30,500	2	6,000	3	10,500	1
Health	37,400	13	614,000	33	15,000	7	120,000	12
Education	27,500	10	210,500	11	11,750	5	130,500	13
Cloth	61,000	21	60,000	3	22,200	10	41,000	4
Farm labour and tools	40,650	14	197,400	11	0	0	253,450	25
Others	0	0	0	0	0	0	61,000	6
Total	286,250		1,850,550		216,810		1,025,850	
Balance	−6,850		−1,107,950		52,340		101,550	

Table 4.3 (continued)

(b) Edeke (2005)

Income

	E1	%	E2	%	E3	%	E4	%
Crop sales	857,000	94	410,000	83	494,400	76	685,000	91
Livestock/fish sales	40,000	4	60,000	12	80,000	12	40,000	5
Tree crop sales	0	0	5,200	1	0	0	0	0
Rent	2,500	0	0	0	0	0	0	0
Paid employment/labour	14,400	2	3,600	1	78,000	12	0	0
Others	0	0	15,000	3	0	0	30,000	4
Total	913,900		493,800		652,400		755,000	

Expenditure

	E1	%	E2	%	E3	%	E4	%
Food	443,000	38	270,000	30	115,200	19	75,200	9
Religious fees	2,200	0	8,000	1	4,000	1	2,000	0
Transport/maintenance/repair/building	83,000	7	163,500	18	38,000	6	158,000	18
Community fees	5,000	0	6,200	1	2,000	0	4,000	0
Health	40,000	3	16,000	2	69,000	12	40,000	5
Education	120,000	10	25,000	3	8,700	1	80,000	9
Cloth	40,000	3	20,000	2	38,000	6	70,000	8
Farm labour and tools	400,000	34	350,000	39	300,000	50	390,000	45
Others	40,000	3	48,000	5	20,000	3	38,000	4
Total	1,173,200		906,700		594,900		857,200	
Balance	−259,300		−412,900		57,500		−102,200	

much as 100 %. Debt was known to be a particular problem in Edeke. With the exception of M2 in Ekwuloko who was facing ongoing hospital bills due to ill health in the family and had probably borrowed money to ensure treatment, the HHs in Ekwuloko were not in debt. The small negative balance for M1 was more likely to indicate a small positive balance once allowance was made for under/over reporting.

The second point of interest in Table 4.3 relates to the source of income in Ekwuloko and Edeke. HH income in Ekwuloko was spread among a range of sources and crop sales represented between 30 and 50 % of total income. The rest was largely made up from land rental and paid employment as well as sales of tree products and livestock. This illustrates the importance of non-agricultural income in Ekwuloko; although it was difficult to extrapolate for the whole village, interviews suggested that this balance was broadly the case for the area. In Edeke crop sales (mostly yam) comprised 70–95 % of total income of the four HH, with income from fishing and livestock also significant (4–12 %). The income from tree crop sales and rent were of little importance, which was generally true with regard to paid employment with the notable exception of E3.

For expenditures there were also differences between the two villages. In Ekwuloko the percentage spent on food (11–22 %), religious fees (4–13 %), community fees (1–4 %), education (5–13 %) were comparable. Expenditures on transport, building and maintenance were particularly high for the two Igbo HHs (especially M3). For M4 the largest single sink of expenditure was farm labour and tools. HHs M2 and M3 rented land from M1 at the rate of N1,500 per area per annum. M2 was also renting land from M4. In Edeke the two largest expenditures tended to be food (9–38 %) and farm labour (34–50 %). This appeared to be contradictory. After all, one would expect that because these were farming HH spending much of their income on hired labour, it would not have been necessary for such a relatively high expenditure on food. However, it can be explained by the relatively narrow crop base of Edeke noted earlier, and especially their inability to grow certain vegetables. The four farmers were growing their crops largely for revenue and not for home consumption. This was especially the case for yam, the crop that formed the mainstay of the production and income (some 87–97 % of total crop income is from yam). Expenditure on building, maintenance and transport for the four Edeke HHs was lower than that of Ekwuloko at between 6 and 18 %. This was not surprising given that all four HHs were based in Edeke rather than earning salaries from outside the village and none had many buildings to maintain. While comparisons were problematic with some of the lower figures (i.e. those below 10 %) it would appear that proportional expenditures on health and education were comparable across the two villages once the especial case of M2 was taken out of the equation. It also appeared as if the expenditures on community and religious fees were significantly lower for the four HH in Edeke relative to those of Ekwuloko,

Income for the two Igala HHs (M1 and M4) was dominated by rental of land and income from sales of crop, livestock and tree products. For M1 86 % of income came from these sources while for M4 the figure was even higher at 94 %. For the two Igbo HHs, (M2 and M3) income was far less dominated by rental and sales of farm products. The figures were 56 and 33 % for M2 and M3 respectively. For these two HHs paid employment was important (44 and 67 % of total income).

Expenditure was relatively high for M2 and M4. For M2 expenditure on health care was particularly high as one of his daughters was seriously ill in 2004—hence the expenditure of N614,000 for the year and a net deficit in the HH budget. For all HH the percentage spent on food (11–22 %), religious fees (4–13 %), community fees (1–4 %), education (5–13 %) were comparable. Expenditure on transport, building and maintenance were particularly high for the two Igbo HH; especially for M3. For M4 the largest single sink of expenditure was farm labour and tools. Neither M2 nor M3 mentioned rent payments, although the land they farm was not theirs. HHs M2 and M3 rent land from M1 at the rate of N1,500 per area per annum. M2 was also renting land from M4.

The difference between income and expenditure suggested that the two were broadly in balance, although HH M2 clearly had problems related to the sickness of one of the daughters. This illustrated how medical treatment can seriously upset the HH budget and indeed lead rapidly to debt if the balance had to be borrowed. However, in general there were few surprises here regarding the main sources of income and expenditure sinks. DDS had long been aware of the problems of education and health care as serious expenditure sinks, and had been involved in various projects over the years designed to help mitigate some of this. The budgets added further confirmation that M4 was the wealthiest of the HHs and, allowing for the unusual circumstances of M2, M1 was the poorest.

4.9 Vulnerability and Institutional Contexts

The sections above have already described some of the trends and shocks found in the two villages, much of which was already familiar to DDS but there were also surprises. Increasing population in Igalaland continues to put more pressure on land, although younger men in particular tend to retreat from farming and migrate in pursuit of paid employment. Fallow periods are generally in decline and crops such as cassava which do well on depleted soil are increasing in acreage. Flooding, while positive if it happens at the expected time, can be damaging when crops are newly sown and can also result in a loss of property and even death. Added to this was the more macro-scale uncertainty generated by the economic and political situation in Nigeria at the time of the SLAs, with local government workers, teachers, pensioners and many others often not paid for months. Thus it was certainly not difficult to imagine the challenges faced by HHs in villages like Ekwuloko and Edeke. This tapestry is common not just in Igalaland and Nigeria but throughout West Africa and beyond. However, the two villages presented some interesting variations on the broad theme.

Ekwuloko is a village of two communities—Igala and Igbo. The Igbos are immigrants, even if their place of birth is only a few miles away (the Igboland border is within a few miles of Ekwuloko). At the time of the SLA there seemed to

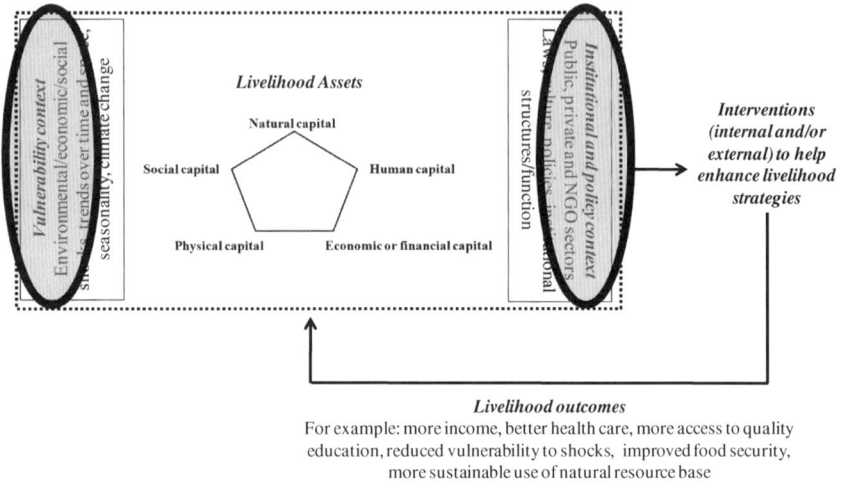

Fig. 4.21 Vulnerability context for the capitals

be an abundance of land in Ekwuloko which explained why Igbos migrated there. Thus the two populations had different asset bases. The Igalas had access to more land which they owned while the Igbos rented. The HHs were generalists who grew a variety of crops and trees and engaged in off-farm activities, including paid employment. The less diverse cropping system in Edeke with a dominance of rice and yam led to greater specialisation and as yam was an expensive crop to produce an elaborate system of seed supply and credit grew along with it, generating peaks and troughs for availability of finance throughout the year. The HH budgets in Edeke painted a picture of credit dependency, and those HHs also had a smaller asset base that those in Ekwuloko. The Edeke HH rented their land and needed access to credit and good quality seed yam planting material.

Social networks appeared to be more diverse in Edeke where HHs had a higher degree of membership of such groups. Agriculture-based groups seemed to be especially strong in Edeke. In theory, such networks should provide a valuable support-base for livelihood. Information could be exchanged and as some of the groups were state or national in scale they could provide a valuable basis for feeding into policy. There was no evidence to suggest that such groups were under threat and, if anything, their diversity and scope seemed to have increased rather than diminished. The extent to which they provided real support to livelihoods was difficult to say, but it seemed reasonable to suggest that membership was advantageous and the diversity of the groups seen in both villages was a positive contribution to livelihood.

So how do these differences affect vulnerability to shocks? This is a difficult question to answer despite the volume of information collected. In Edeke the livelihood base was more specialised (narrower) than Ekwuloko suggesting that they were more vulnerable to shocks. A catastrophic flood event would cause extensive damage and potentially wreck livelihoods at least for a time, and indeed this has been the case in 2012 when the River Niger burst its banks and caused extensive damage to the farms and property of riverine communities. Given the changes being brought about by global warming, it is possible that river levels could rise resulting in floods that could increase in both frequency and intensity. Also, it is not just the natural shocks which need to be considered. The reliance on but a few crops for the bulk of HH income puts them at the mercy of price fluctuations. The price of yam has remained relatively high as many farmers in Igalaland and elsewhere find it increasingly difficult to cultivate and have tended to switch to cassava. The yam specialists of Edeke could do well in these circumstances especially as yam remains a highly coveted food, but this too could change. Did they have other options for replacing yam if prices crashed? Given the nature of their land the answer was probably no. Other crops could be grown but they would not be as financially lucrative as yam, and fishing was also not a good substitute. One approach would have been for them to reduce the costs of growing yam, and one of the major costs was seed material which they import each year from specialist seed yam growers along the western bank of the Niger River; but this is not an easy option and credit is critical. Indeed it is worthy of note that the Edeke community has had a long association with DDS and at least in part this is because of the micro-credit scheme. The dominance of the river in their lives is indeed a mixed blessing. It 'gives' in terms of land replenishment, income from fishing and ease of transport but it can 'take' because of its ferocity. At the time of writing the 2012 flood, the worst for 40 years, has resulted in many communities who reside along the banks of the Niger having to migrate to Idah. They have lost much of the crop produce and the impact on their livelihood has been severe.

The wider base of livelihood in Ekwuloko suggested a greater ability to deal with shocks and that village had certainly survived for many years. Its links to the Igala ruling families also seemed to add to a greater sense of resilience. The hybridisation of Igala and Igbo cultures provided a unique sense of looking towards both Igalaland and the very much larger Igbo populations of the South East of Nigeria. Transportation links have been improved with the construction of a major highway linking Idah with Nsukka and on to Enugu. The cropping system is diverse relative to Edeke and drops in some crop prices can readily be accommodated by a shift to other crops or indeed other forms of income. Despite all this the problems faced by Ekwuloko were probably the same as elsewhere in Igalaland—primarily the danger of losing its younger generations as they sought gainful employment elsewhere. They may not face the same threat of catastrophe as do the HHs in Edeke but there are challenges to be faced nonetheless.

In terms of institutions in both villages there is a virtual absence of government (local, state and Federal) structures or interventions other than schools and health clinics. Ekwuloko only has one primary school and no secondary school. Edeke is better served, with four primary schools and three secondary schools, but as in Ekwuloko as soon as the youth leave school they often relocate to the cities. In Ekwuloko the presence of a State-Government managed Forest Reserve some 3–4 miles from the village provided a resource for some farmers to utilise for crop production, including yam, even if this is illegal. It should be noted that none of the Ekwuloko HH included in the SLA were involved in such utilisation of the forest reserve. The main Nigerian political parties have a presence in both villages although it has to be said that they do not appear to be a positive force for enhancing livelihood. Both villages also have a number of Churches and their positive role in terms of social capital has already been noted. To what extent these help is not known other than a sense amongst the HHs that they are positive. The only faith-based group that provides support, at least for Edeke, is DDS. At the time of writing DDS has not established any groups in Ekwuloko but the reasons for that are mixed and not necessarily a reflection of any lack of desire from either party. Thus the institutional contexts in both places do not seem to be especially important with regard to supporting HH income generation and livelihood, at least in any direct sense. Having said that, education is, of course, important and the presence of schools that are accessible to the respective communities cannot be underplayed.

4.10   Did SLA Succeed?

The final section of the chapter will draw together the components of the SLA picture and evaluate the advantages and disadvantages of the approach; it will outline the learning gained by DDS to inform changes to its credit scheme. Did the analysis succeed in producing a story that DDS could use as a basis for change?

It must be noted that any SLA is by definition unique to the specific context within which it is applied. Thus there are facets of the Ekwuloko and Edeke analyses which only apply to those places. Ekwuloko is a border village in every sense of the term. It is close to the border between Kogi and Enugu States but more importantly lies within an area of ethnic mixing. Many villages in that region have the same mix of Igala and Igbo and to the west of Igalaland borders with Benue State has villages with mixed Igala and Idoma communities. Nigeria is a country of 160 million people with hundreds of ethnic groups. Social borders between these groups are fuzzy. Thus while on the one hand any attempt to generalise the findings from Ekwuloko village may well be resisted this should not be taken too far. Similarly, the village of Edeke is representative of thousands of villages along the banks of that great Niger but are obviously far different in context from the hundreds of thousands of villages that exist away from the two main rivers. Dismissal of any attempt to generalise from the Edeke results is all too tempting and to some extent warranted, care therefore needs to be taken to avoid rejecting any wider lessons that can be gleaned.

Nonetheless the case-study foundation of much of the SLA literature can be problematic precisely because it is so easily dismissed as being 'site specific'. While it is 'analysis' in a real sense such studies can also be labelled with that most deadly of labels—being descriptive. Ironically this label is also often applied to SLA studies in general (van Dillen 2002). In terms of the central import of SLA as a means of bringing about change this frankly does not matter, but there is a case to be made for enhancing the potential for comparative research with SLA so as to identify patterns (de Haan 2005). In this study an attempt was made to compare two villages that were expected to be quite different but this was not so much to identify a pattern, not possible with just two places in any case, but for DDS to learn about two extremes and how a new credit scheme may apply. While a more meta-SLA approach may be attractive in theory the practice may not be easy. First there are issues of commonality in approach (or lack of it). SLA is a broad banner that covers many disparate practices even if the underlying philosophy is constant, and this may not help comparisons. Secondly, there is a 'lower common denominator' issue. Some SLAs are employed in greater depth than are others, thus the additional knowledge from some SLAs may be wasted as comparisons are not possible with other places where that knowledge was not gleaned. Therefore the danger is one of only being able to identify somewhat large-scale patterns that are almost meaningless; for example, that 'agriculture is an important component of livelihood in many places'!

Similar issues over representation arise within villages. Only four HHs were assessed in each place and they obviously represent a small proportion of the village populations. There was an attempt in Ekwuloko to locate the four HH within the population using some socio-economic indicators and results suggest that the four HHs were at the upper end (better off) of the spectrum. In Edeke it was not possible to do this for logistical reasons. Thus even with a better spread of HHs within the spectrum in Ekwuloko, there is always the criticism that those selected are unrepresentative of the village population as a whole. Community is indeed a myth and populations are not homogeneous. Conclusions drawn from the SLA on the four HHs in Ekwuloko can readily be contradicted by talking with members of another HH in a different compound on the same road. If one is only attempting to help those four HHs then it may not matter, but if the aim is to draw out generalities that apply to more HHs then diversity will always win and some will either not be affected at all by any planned intervention and some could even be disadvantaged.

What do the SLAs say about the two villages? It is clear that Edeke is a quite different agro-ecological and socio-economic environment to Ekwuloko but that is not surprising as Edeke was chosen with precisely that in mind. It would have been astonishing if the results for the two villages were similar. Credit was clearly an important issue in Edeke and less so in Ekwuloko, and in part this explains the longer history of engagement of Edeke with DDS. Some of the female (and male) money lenders reside outside of Edeke and wield much power. In a focus group discussion one of the four HHHs said that he was wary of upsetting the money lenders by not borrowing from them in any year as he may require credit the following year. Breaking HH dependence on such credit traps is not easy given that whatever is done has to ensure sustainability; not availing of such credit facilities in one year means farmers have almost to go on their knees to obtain it in subsequent years. Confidence is a critical ingredient for sustainable livelihood.

Can it be said that Ekwuloko is more sustainable than Edeke? Asking this question and indeed attempting to answer it is certainly tempting, but can it be done? At one level it may seem that Ekwuloko is the more sustainable of the two as it has greater diversity in livelihood, at least for the HHs that were assessed. However the Edeke agricultural system works well even if the HHs have to source expensive yam planting material from Edo State, and there is definite reliance on credit. If sustainability is taken to be the ability of the system to resist and survive shocks then Ekwuloko would arguably be in a better position to switch between livelihood options. What would the HH in Edeke do if credit was no longer available from current sources at a reasonable cost? The problem with such hypothetical questions is that markets can shift. Why would the Edeke HH not be able to source credit from other sources as they have done in the past? It would take a major upheaval, such as rising river levels as a result of climate change, to bring about a major disruption to their way of life and that would also impact on millions of others. Indeed this is precisely what has occurred in 2012. Such a severe 'shock' to Edeke has had a major impact on their livelihood and thus highlights the difference between the two villages. But the severe flood is a one in a 40 year occurrence and provided it does not happen every year then the communities will recover. It is undoubtedly the case that the future year or so will be a tough one for the Edeke people as they have lost so much in 2012, but the soil fertility will still be there. Therefore any attempt to empirically compare Edeke and Ekwuloko by using some normative definition of sustainability is problematic.

With regard to the SLA methodology it has to be stressed that the work involved, even when looking at relatively few households, was substantial. DDS expected this and hence during the process of village selection care was taken to think through the costs and the logistics involved. DDS's prior knowledge of village HHs, in both places, but especially Edeke, was of immense help and findings could be checked constantly with those who knew both villages well. DDS has many key informants in Igalaland including the Attah of Igala (the Chief of all Igala people) and from among the District and Village Chiefs, civil servants, business men and women, religious groups (Christian and Muslim) and village folk. Thus findings can be readily checked. Trust in DDS staff was always evident amongst the households. But even with all these advantages and a relatively narrow focus on a few households the SLA was far from being easy or cheap. At its height, DDS had one member of staff dedicated almost full-time to the two SLAs (a total of two years) and sometimes up to four other staff were drafted into help. Add to this, the cost of transport and subsistence for field work as well as data processing, costs are significant.

As important as the logistics are, even more important is that to be effective an externally-led SLA depends upon the quality and quantity of information supplied by participants. Indeed one of the criticisms often levelled at SLA is that it ends up being a cataloguing exercise which generates long lists of figures which can lessen the priority of 'people'. Trust is an important element and so is the need to 'truth' information which HH's supply. In both cases, but especially in Ekwuloko, there was a tendency for respondents to downplay their ownership of assets. In the case of farmer M4, for example, who admitted he did not mention in early

interviews, a larger tract of land he owned. This, of course, would have a significant impact on an assessment of his asset base and hence options for improving
livelihood sustainability. There are various reasons why HHs would want to do
this. For example:

1. *Tax* (or other government intervention). A fear that those engaged in SLA
   would report the findings back (directly or indirectly) to government officials
   who may then seek to increase taxes.
2. *Theft*. Revealing the ownership of assets may make the HH a target for
   thieves as armed robbery in Nigeria was widespread and with the opening up
   of Igalaland with new roads this had even spread to areas that had been relatively free of armed robbery. Given that Ekwuloko was more vulnerable in this
   sense, and more accessible to vehicles, this was more of an issue there than in
   Edeke.
3. *Development*. Claiming that a HH has less assets than they actually own could
   be seen as a means of enhancing the chances of more development coming to the
   village, or more precisely to the HH. The perception here might be that claiming
   relative wealth would be detrimental; any resources would go elsewhere.

Thus it should be noted that HH do have motives for not providing information or for exaggerating or downplaying some facts. The assumption that people
will always tell the truth is wrong, but they have good reasons (at least from their
perspective) for being economical with the truth. This does not mean, of course,
that skilled SLA facilitators can compensate to some extent for this. Triangulation
of methods can help a lot. For example, observation can be used alongside formal interviews and more informal discussions. It was clear to the DDS staff
engaged in the Ekwuloko SLA that farmer M4 was 'hiding something' and thus it
was possible to tease out from him at least some of the additional land he owned.
Observation can go a long way to identifying at least the physical assets and
this process can in turn encourage better cooperation as respondents realise that
some aspects of HH livelihood cannot be hidden or exaggerated. Thus the benefits of cataloguing assets were not simply the numerical lists generated at the end,
although there are limits to what even the most skilled external observer can glean.

A second issue revolves around exactly how the SLA was employed to help
inform the activities of DDS. The SLA model as suggested by DFID implies two
forms of intervention which could take place (Fig. 4.22). Firstly the SLA could be
part of a participatory process to allow HHs and the 'community' to learn about
themselves and how they can best overcome obstacles or take advantage of opportunities that may come to light. The limitation here, is the assumption that the SLA will
bring to light these aspects of which HHs were not previously aware. The danger is a
simple response of 'well we know that already so why did we have to go through all
this?' or perhaps even 'well OK but we don't have the necessary resources to address
the problems that have been highlighted'. More often than not, the SLA is a prelude
to a planned intervention on the part of external agency acting on behalf of the community. Thus the SLA may generate learning at the local level but the findings are
used to plan interventions which can help the community. This is illustrated by the

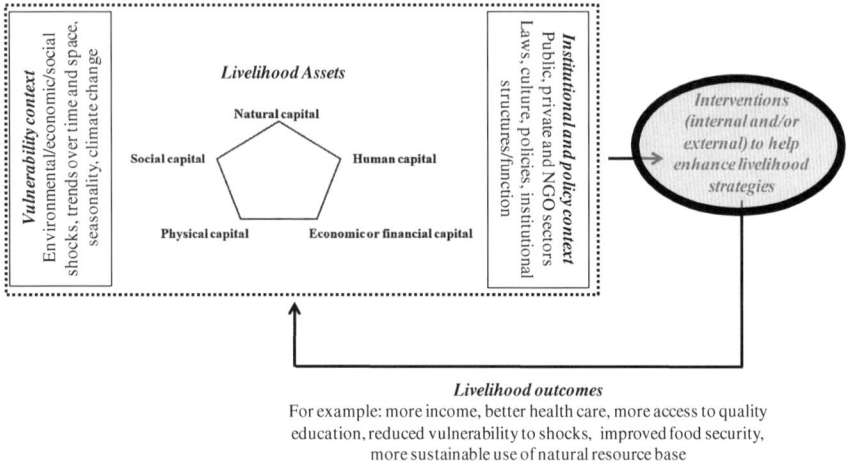

Fig. 4.22 Interventions

boxes and arrows to the right of the DFID diagram. There is obviously a need to consider wider policy and other contexts and constraints that may operate.

Interestingly, in the case of DDS, their motive for the SLAs in Ekwuloko and
Edeke was to seek evidence to support an assumption that their credit scheme
needed to be re-modelled and redirected and were exploring ways of how best to
do this. They were especially keen to move towards a more specialised approach
suitable to current situations. Given that the SLAs were funded in part by DFID
as a component of a seed yam project, and credit was likely to be a strand of the
strategy, DDS saw the SLAs as providing useful information as to the range of
income generation with which HHs could engage. But why go to the trouble of
such detailed analyses? Why not rely on existing knowledge of the local context?
DDS certainly had much local knowledge and experience gained over 40 years
of working in Igalaland, including the running of saving and credit schemes for
much of that time. At various points in the chapter it has been noted that some
of the findings coming out of the SLA were familiar to DDS, and given that the
organisation is staffed by Igalas and Igbos that should not be all that surprising.
However it felt bound to take a fresh look at changing conditions locally, nationally and internationally with any changes and adaptations connected to evidence.
It was also the case that DDS gained new knowledge, especially with regard
to Ekwuloko. Thus the SLAs were a part of an attempt to overhaul and evaluate existing interventions rather than about creating new ones. Despite the presence of the familiar, DDS certainly realised that it learned a great deal from the
SLAs and its new credit scheme, Farmers Economic Enterprise Development
(FEED), does have a much stronger emphasis on recipients showing how credit
helps enhance their livelihoods. Pining down a cause-effect relationship between
the SLA and FEED is not easy given that DDS went into the SLA process with

an underlying 'feel' that their credit scheme needed to be overhauled and better linked to investment. In fact the growth of the new approach to credit occurred in parallel with the SLA rather than at the end. Reliance on credit in Edeke was especially pertinent and DDS has already reviewed its credit provision to the riverine communities; the matter is complicated given the presence of many local credit providers.

Nonetheless the question has to be asked as to whether a more 'quick and dirty' approach would have succeeded better than what was achieved by DDS? Frankly while this would have been a great deal cheaper and quicker it is doubtful whether it would have been any more successful in these circumstances and may well have been less so. Indeed did DDS need to do an SLA at all given their extensive knowledge of Igalaland? Could FEED have been designed without resorting to SLA? The fact that DFID was providing some funding for the SLAs may well have influenced the decision of DDS to widen the remit of the work, but such 'what if' scenarios are not easy to dissect. DDS senior staff were certainly adamant that its redesign of the credit scheme would be based on "village level studies" as to what is required irrespective of DFID's involvement. To some extent, the fact that DDS participated in the SLAs as a basis for redesigning its credit scheme does provide it with much credibility as such thorough research in finding a base line for FEED is impressive, especially for partners upon whom DDS obtains much of its funding. These constantly require DDS to provide arguments and evidence for its planned interventions; at that point DDS was seeking support from an Irish aid agency to provide a 'kick start' for FEED and of course against such a background of research this assistance was forthcoming without further questions. Indeed herein rests an answer to one of the dilemmas inherent in the SLA namely how is 'success' to be assessed? Is it with the quality of the analysis, whatever that may mean, or quality of the analysis as a trade off with cost? Alternatively should success be gauged in terms of any change which followed the SLA rather than the SLA itself? These are not unrelated and presumably all three of these stances could be adopted, albeit with greater emphasis upon what was finally achieved. DDS certainly saw the SLAs in Ekwuloko and Edeke as a success, and the farmers involved, especially those in Edeke, continue as DDS partners. At the time of writing (2012) DDS maintains that the FEED programme is working well, although they have not as yet extended it to Ekwuloko. The reasons for the latter do not appear to be related to any breakdown in relationship, but more the result of a number of factors including a change in leadership at DDS, a temporary shift of attention and resource towards some other projects that the organisation has been engaged in and a reduction in the number of personnel. It is not yet possible to talk about how the credit scheme has impacted upon HH livelihoods and which in turn enhanced (or not) the asset base. Hence the final step of the SLA—showing a positive impact on livelihood of the participating households—has not been achieved in a formal sense (Fig. 4.23) and may be a significant challenge given that the changes were being planned prior to and in parallel to the SLA. At this point it may be more logical to think about the impacts upon DDS rather than the households. Opportunities to do this are awaited.

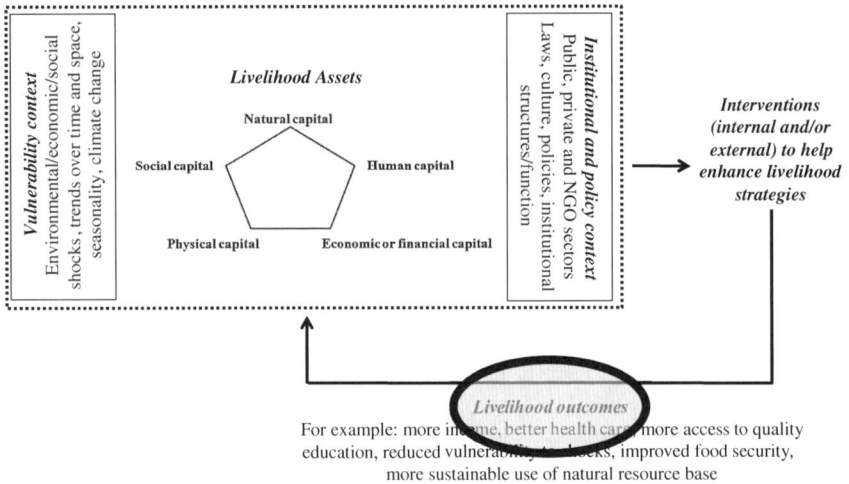

Fig. 4.23 Livelihood outcomes

4.11 Conclusions

Compared to the studies reported in the literature the SLA conducted by DDS was unusual in a number of respects. The motivation was clearly focussed upon providing:

• A backdrop of HH livelihood which would allow the results from the seed yam trials to be seen in context
• Information on livelihoods which could be used as a 'feed' into the redesign of the DDS credit scheme and its presentation to potential partners for support
• An introduction into a new and important location (Ekwuloko) where DDS had only little previous involvement.

The first of these is perhaps more in tune with how SLA has tended to be applied by researchers and practitioners, at least as reported in the literature. The second and especially the third are novel uses of SLA as far as these authors are aware. The literature on microcredit is a substantial one and individual systems such as that of DDS are known to evolve as circumstances change. In itself that is not a new insight. But the use of SLA to feed into the design of a microcredit scheme appears to be new. The third use of SLA listed above is perhaps the most unusual of the three. SLA can certainly provide a means by which a community such as Ekwuloko can become 'known' to an outside agency, especially as applied here by DDS over a long term and in quite an intimate fashion. Much time was devoted to the interaction by DDS and the selected HH in Ekwuloko and it must be remembered that one of the factors in selecting the four HH was their leadership roles in the Igala and Igbo communities of the village. This use of SLA as a means of 'getting to know' each other does have a logic, especially when seen in conjunction with the other reasons and the resources provided by DFID and others that helped to underpin the SLA. Whether an SLA would have been employed solely for that purpose is doubtful. However, it is important to stress the caveat that much of the SLA experience has perhaps not been documented, at least in a readily accessible form, so it may be that parallels to the uses made of SLA by DDS exist elsewhere.

Resources were provided by DFID, via the seed yam project, and Gorta who sponsored the credit scheme. These are inter-related, given that credit is important for the sustainability of the seed yam system being introduced to farmers. Resources are always limited so inevitably DDS had to engage in some trade-offs between what it would ideally like to do as set out by the SLA framework discussed in Chap. 2 and what was feasible. The result of this was a decision to focus on an SLA of the HH engaged in the seed yam trials; even if this this meant that the numbers were relatively low. From the perspective of DDS this focus delivered what they wanted to achieve and at a cost that could be covered. Success was measured in terms of meeting those objectives and it managed to do just that. There were no academic publications in mind when the process began except for an 'output' (report) required by DFID for the seed yam project. The SLA was not intended as a basis for further research.

The broad nature of the intended intervention (i.e. microcredit) by DDS was known before the SLA began. In theory the SLA demands an open mind with regard to strategies; one has to analyse the situation and base any interventions upon what is found. An open mind should be kept as the SLA can suggest interventions not immediately obvious to those implementing the process. Indeed the SLA may suggest that little constructive work can be done in the circumstances, or perhaps it may result in a need for further research. In this case DDS were in the process of making changes to their credit scheme across their membership and were using the SLA as a means of informing the change. Hence while the SLA may have been broad in its scope it wasn't geared towards looking for suitable strategies in an 'open minded' sense. However, it has to be said that this is not especially unusual for SLA. Many of the examples covered in Chap. 2 involved the use of an SLA linked to an already identified intervention or perhaps to explore options regarding a specific problem. For example, Wlokas (2011) used SLA to explore potential of solar water heaters for households in South Africa and Cherni and Hill (2009) used the approach to explore energy supply in Cuba. Thus it is understandable that examples of 'blank page' SLA may be rare and its link to an already established intervention or a discrete problem may be more the norm.

Thirdly the process was a long one. The costs were controlled to match what DDS thought was acceptable, but each of the two SLAs still took a year to complete. This was partly because of the need to have specific data during the whole growing season-market cycle. While donors and researchers may not consider the effort required in doing an SLA there is no doubt that DDS put quality time and energy into what they did in Ekwuloko and Edeke. Their constant visits, observation and interviews demanded significant time on the part of the DDS staff involved, this was compensated by the depth and breadth of the knowledge acquired of the villages and its peoples; a point that was of especial relevance for Ekwuloko. Based upon the SLA literature this time put into the process by DDS would appear to be on the long side, especially given the small sample size in each of the two villages. Admittedly this is not often reported in the literature so it can be hard to gauge. As noted in Chap. 2, sample sizes for reported SLAs can be in the hundreds and thus while they may take months to complete the time allocated for each respondent may be relatively small. The DDS case study effectively involved an equivalent time allocation of 3 months (=12 months/4 HH) for each HH. This might not sound like much time to try and understand the livelihood of a HH, and indeed it is not, but is significantly longer than many of the reported SLAs in the literature would have allocated.

SLA in itself does not avoid some key issues in deriving an evidence base for action. As with social science research in general, the answers (evidence) that one gleans is influenced by the ways in which that information is collected and analysed. While it is broad in scope SLA is certainly not a panacea that avoids these issues. After all, it is still largely an approach implemented by 'outsiders' to the community whose livelihood is being explored and limits on time and resources inevitably force a compromise as to the depth of exploration. Community participation may be nothing more than an extractive process to help fine-tune the external vision of sustainable livelihood and a means of providing information. As a result there are some real dangers here, clearly outlined by the post-development movement. Even with the very best of intentions it is still likely that the SLA practitioners will miss much in the search for relatively 'quick' and 'representative' (as seen by the practitioners) answers. The Igala experience of SLA was idyllic as conditions conducive to its success abounded. Loop holes are always possible although even these will be valuable as part of an on-going learning experience.

Chapter 5
Livelihood into Lifestyle

5.1 Introduction

In the previous two chapters the authors provided an in-depth example of an SLA in practice. The context, at least on the surface, was a familiar one for SLA; a predominantly rural environment in a developing country, Nigeria. The case study revolves around the activities of an NGO—the Diocesan Development Services (DDS)—and as with all SLAs there are various contexts which are important and these have been summarised in Chap. 3. There was a perceived need by DDS to understand livelihoods in two Igala villages (Ekwoloko and Edeke) and the SLA followed the framework set out in Chap. 2. Admittedly this case study-based approach is prone to the criticisms that normally surround case studies in general, most notably the wider applicability of any lessons that may arise out of the analysis. But while there is much to the DDS-SLA story that is familiar there are also some interesting and unique features. Indeed the intention was to bring together insights from the existing literature on SLA and the lessons that could be gleaned from its application in the case study; an approach that was termed 'phronesis' (a form of practical wisdom) by Aristotle.

The perceived need for the SLA came from reflections within DDS and was not an external requirement imposed upon it by outside partners. Neither was it financed or planned by an agency outside of Nigeria or indeed Igalaland; on the contrary, it was decidedly a local SLA and does provide some contrast with the broad tendency in the SLA literature which implies that the SLAs were largely driven by outside agencies. In fairness this criticism is not easy to discern from the literature as the reasons for the SLA in relation to who decided to do them and why are often missing or unclear. Typically the literature gives the sense that the SLA is an early stage within an externally funded research project; designed and implemented by groups living some distance from the local communities who were the focus. DDS's intentions for engaging with the SLA were multi-faceted and were to help:

1. Provide an input into the reorganisation of the microcredit scheme that had been in place for some 30 years.

S. Morse and N. McNamara, *Sustainable Livelihood Approach*,
DOI: 10.1007/978-94-007-6268-8_5,
© Springer Science+Business Media Dordrecht 2013

2. Identify gate-keepers in Ekwuloko village through which DDS could reach out
   to the wider community.
3. Bring overseas donors on board for the proposed changes to the microcredit
   programme.

The first was tangentially related to the involvement of DDS in an on-farm
programme of research funded by DFID and thus has some resonance with the
notion of SLA demands by a distant agency. However, the DFID project did not
require an SLA to be done—the emphasis was upon the agronomic and economic
aspects of the on-farm trials with seed yam—but DDS saw an opportunity for a
degree of piggy backing the cost of the SLA onto the trials. This seemed fair as
DDS staff were involved in the trials and had to travel to the villages on a regular basis so why not avail of that opportunity to expand the potential for further
insight? Indeed, while the DFID project did not require an SLA the whole outcome would be enhanced by putting the results from the healthy seed yam trials
into a wider context. In turn, the DFID project was keen for DDS to sustain its
support for the healthy seed yam system by providing microcredit. The intervention offered a mix of opportunity plus restraint; opportunity by using the DFID
project to support the SLA; restraint because resources were limited to allowing
for anything wider than a focus on the few households in each village.

The second point was motivated by deliberations within DDS to extend its
activities into a 'new' area. As highlighted in Chap. 3, villages in the border areas
between Igalaland and Igboland tend to have characteristics that are different from
where DDS has functioned to date; Ekwoloko in particular provided an opportunity to establish a bridgehead of activity that could be expanded upon. Border
areas are important areas for a variety of reasons, also covered in Chap. 3, and
the need on the part of DDS to include them is understandable. In effect the SLAs
were almost a 'getting to know you' exercise which could help build trust; an
important factor also in social inclusion as people in the border areas now had an
opportunity to study again what opportunities DDS could offer them.

Thirdly the SLA was seen as being of value in terms of DDS making a case to
overseas partners for support with the microcredit scheme. The intended revamp-
ing of the microcredit scheme was towards a more business-plan approach that
mirrored what formal bank lenders were already doing. This required borrowers
being asked to set out in detail how they intended using credit, what costs and
revenues they might expect and challenges they may face. The farmers involved
in the on-farm trials were engaged in a pilot study of this process and DDS felt
it would help improve their case for funding when they were able to provide a
complete picture of the process in which the farmers were involved. DDS is no
different from other development NGOs in the Global South in seeking partnerships with donors for development programmes. This process requires constant
interaction and understandably partners need to see evidence of engagement and
impact from those they fund. This evidence is obtained in many forms including
reporting and field visits, and any new project has to be justified. Understandably
agencies such as DDS have always engaged their partners in dialogue which

included a genuine understanding and appreciation of problems and achievements. Reciprocity included hard comments at times but on balance relationships were understood to be part of the healing and restorative justice that is part of the total development process. Why is *development* required in the first place? The imbalance that gave rise to it dates back over half a millennium and hence the need to tackle the process with much respect recognising the dignity of each participant. The value of reciprocal partnerships is by no means a new insight or indeed one restricted to Catholic Church-based agencies, the literature on donor-field agency relationships is an old and extensive one. For a flavour of main issues the interested reader is referred to a series of papers written by the authors that have explored relationships between a number of Catholic Church-based field agencies in Nigeria and development donors from four countries (Morse and McNamara 2006, 2008, 2009).

It also needs to be noted that DDS had arguably applied SLA in the sense of it being a set of guiding principles for action since its inception in the early 1970s. It had spent many years working with local communities, getting to know them (and the communities getting to know DDS) during which there was an implicit assessment of the important facets within SLA such as capital, resilience and institutions as a prelude to intervention. Such terms were not in vogue then but the concepts existed and were well understood and practised. The more formal SLA structure was not known but in essence its framework and guidelines were implicit in much of what was done. Indeed the undeniable logic of the SLA is such that most development agencies embedded within a community over a relatively long period will have developed much of what is in the SLA framework.

With these considerations in mind, the decisions by DDS to implement an SLA and to do so in a certain way were influenced by a number of circumstances. The intention was certainly not to do it for academic publication or to influence broader policy at national or local levels in Nigeria and beyond. The lessons to be gained were for internal consumption initially at least. This is within the spirit of the SLA although no doubt the reader can question the decisions that were made. For example, and perhaps most obviously, the small number of households involved in the SLA in the two villages can be readily criticised. Ideally, the sample size should have been greater in order to provide a better representation of the populations, but DDS was willing to trade this desirability for representation against other demands. Hence the typical balance outlined in Chap. 2 where the emphasis with SLA is so often upon more technical aspects of doing it 'right' with perhaps less concern about eventual use of the information is reversed. These are not mutually incompatible concerns; doing an SLA 'right' is fully compatible with making sure that the knowledge is 'used'. But in the DDS example there were trade-offs between these two, driven in part by cost but also by the other motives for doing the SLA.

Perhaps surprisingly, at least to these authors, the reason(s) for 'doing' an SLA in the first place have seldom been discussed in the literature other than in a broad sense of it being necessary to discover what people are doing (as a feed into a rather vaguely, if at all, defined 'policy' or decision-making process) or what impacts an existing intervention might have had or is likely to have. Rarely if ever is there any attempt within the published literature to link the SLA to what has subsequently happened to the communities; SLA journal papers usually end by making recommendations and suggestions for further research or intervention. In other words, did the SLA make any meaningful difference to their livelihoods? Given that SLA is meant to provide a basis for action then this serious omission as to what action subsequently took place and what impact it had is glaring. In part it is perhaps understandable given that the academic literature tends to prioritise papers that are highly focussed, with clear objectives, methodology and results to match. Whether the research has any impact or use is often of lesser, if any, importance. What matters is whether the research underpinning the paper is pushing forward the frontier of human knowledge. As mentioned in Chap. 4, DDS did feel that the SLAs were beneficial with regard to its objectives set out above, but at the time of writing Ekwoloko still remains largely outside of the new microfinance scheme. Thus 'success' is a matter of perspective; what is meant by success and success for whom?

SLA was developed to be applied within the context of intentional development. Thus it embodies, almost by accident, the polarity of developed and less-developed which is often applied, albeit simplistically, to all nations of the world. But the principles upon which the SLA framework is founded can apply to any community on the globe; after all every person has a livelihood even if they don't have to work all that hard for it. In this case, given such a breadth of relevance then SLA could help inform policies and interventions within the developed world as much as it could within the developing. Indeed this takes us to an interesting transition between sustainable livelihoods and sustainable lifestyles, where the latter is more than just earning a living but also embodies culture, recreation, image and so on.

In this chapter the intention is to broaden the points made about SLA within previous chapters. How can all of the points discussed help with an evolution of the concept? What would such a change look like and what advantages, and issues, would it bring? The chapter will continue with a further exploration of the Global North–South polarity in the development and application of SLA and the implications that arise from this. It will then progress to explore the extension of livelihood to lifestyle and what that means for the traditional framework for SLA set out in Chap. 2 and applied by DDS.

5.2   How SLA?

The SLAs implemented by DDS demanded much effort, and it was clear that in-depth material was required given the need to triangulate findings and build up trust with the households. As a result each of the SLAs took one year of constant engagement to fully grasp a sense of the capitals as well as the institutional and resilience contexts. It was, in short, a non-trivial exercise. DDS had the luxury of being able to take this time given that the on-farm trials also took almost a year to complete. In effect the SLAs were implemented as case studies, where a case study can be defined an in-depth investigation of an individual or a group (Armstrong 2006). In the previous chapter it was questioned whether a faster version of SLA could have been implemented via observations, participatory sessions and questionnaire-based surveys. The advantages of this would arguably have been better representation and it certainly could have been a lot cheaper. However, on the negative side there would be issues of superficiality and lack of trust leading to untruthful answers. There were certainly numerous indications with the more in-depth approach taken by DDS that households could be economical with the truth or present their situation in different ways, facts that were only discoverable with the intensive process of observation and double-checking that could take place over time. Given the sensitivity which probably surrounds any exploration of livelihood then this reticence is perhaps understandable and should be expected, but it does raise important questions of data quality and subsequent confidence in the knowledge which is created. Is it acceptable to get an SLA partially correct, at least in terms of what may be seen as the key facets of livelihood? Does it really have to be all that accurate? If not then what degree of inaccuracy is acceptable and how can this be balanced with representativeness? In other words is it better to have a larger sample and less depth or a smaller sample and greater depth? These are highly subjective questions, and arguably the important output is as much to arrive at the key features of livelihood and how they can be improved as it is to obtain extensive and detailed datasets.

The use of such in-depth case studies within the social sciences certainly has its adherents as well as critics, and the debate is somewhat polarised. While case studies do allow for depth of insight they can be dismissed as being 'site specific' and hence 'descriptive' in nature (Miller 1977; Forsyth 2006), and thus having little of wider value. Putting aside the reasons why DDS implemented the SLAs and focussing instead on the wider value for informing intervention in Igalaland let alone anywhere else the focus on just eight households in two villages would certainly suggest that the value of any insights is limited. This conclusion does, of course, ignore some of the reasons why DDS implemented the SLAs in the first place, but it does nonetheless seem reasonable. Researchers such as Campbell and Stanley (1966), Dogan and Pelassy (1990) amongst others have provided the case against the use of case studies. But the approach also has its enthusiasts, primarily because it does allow for an in-depth and sustained study of what can be extremely complex dynamics (Forsyth 2006). This means that the findings can have a strong validity as time is allowed for triangulation and checking of insights (Miller 1977), although it has to be said that there is still potential for bias on the part of the researcher; they can still only see what they want to see. Ironically one of the case study enthusiasts was the same Campbell who along with Stanley initially referred to case studies as having no *scientific value* in their 1966 publication (page 6). He completely changed his mind and became one of the case study's strongest proponents (Campbell 1975). Others who have provided strong a strong defence of the case study are Geertz (1995), Flyvbjerg (2006). But even so the debate over the value of case studies still seems to be highly polarised. One way of trying to bridge the gap is to look for patterns with a range of case studies and de Haan (2005) amongst others have suggested with SLA. Given that the SLA framework is broadly consistent across studies then it could potentially allow for a more qualitative 'meta-analysis' of the results to see what patterns may emerge, but to date there has been little research in that direction. It also has to be stressed that while the debates over the value of case studies can be somewhat polarised this is often not an 'either/or' situation and case studies can be used as the prelude for wider surveys and indeed theoretical analysis (Armstrong 2006).

However, as has already been noted care does need to be taken when applying such critiques of the case study approach to the work undertaken by DDS. The object of the exercise from the point of view of DDS was multi-faceted but was not primarily to generate information for presentation within journals and books; the development of this publication being very much an a posteriori event. It was not even intended that the results would be 'scaled up' to drive DDS policy elsewhere in Igalaland. It is true that it was at least in part designed to help inform the development of the microcredit scheme, but there were many other influences at play within that and the nature of the FEED programme (the new form of microcredit adopted by DDS) was already being explored at the time the SLAs were planned. The SLA results were one input into that process and helped illustrate the value of the proposed changes to potential partners. Hence while one can have some sympathy with critiques over the data collection methods within SLA these should not lose sight of other important dimensions inherent within the framework.

5.3 Where SLA?

It is perhaps worth reflecting at this point upon the geographical disparity presented in the SLA literature. At the outset it has to be stressed that there are dangers of over-simplification given the globalised world of ideas in which we live. Ideas development in one place can rapidly evolve and expand elsewhere. Indeed the people who develop ideas can be highly mobile, thus making it difficult to tie ideas down to one place or indeed one time. Looking at Table 2.1 it is clear that SLA originated from a range of ideas and influences from across the globe. Amartya Sen's book *Commodities and Capabilities* was influential, and while Sen has held academic positions in both the USA and UK, he is from India. Indeed his insights, as is so often the case, did not exist in a vacuum but were built from existing knowledge on poverty. Similarly, international agencies such as the UNDP have played a major role in the evolution of the ideas upon which SLA is based. It also has to be repeated that the analysis of the academic literature may not necessarily be a reflection of SLA activity as a whole. Much of the work may not be documented at all or perhaps in formats that are not readily accessible such as internal reports, project proposals and evaluations. Nonetheless a geographical

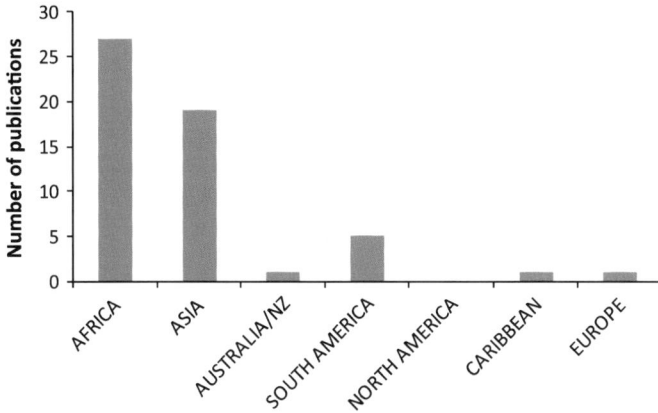

Fig. 5.1 The focal regions of reported SLA's

analysis of who 'does' SLA and where it is largely 'done' is informative, even if based upon a limited sub-set of activity.

Figure 5.1 presents the analysis of the literature in regions where the SLA was implemented. The bars represent regions and the vertical axis is the count of publications from those regions. As the graph shows, the bulk of locations were in Africa and Asia, and indeed of the 46 countries that are specifically named in the publications (rather than just a reference to a region or unspecified) 31 (67 %) are in Commonwealth countries and 15 are not. By way of comparison, it is usually assumed that there are 193 countries in the world (at the time of writing) and the Commonwealth has a membership of 54 countries (28 % of the total number of countries). The population of the world is seven billion people out of which two billion live in Commonwealth countries, equating to 29 % of the total. Thus the emphasis on Commonwealth countries in the SLA literature is remarkable and to some extent understandable given the history of SLA discussed in Chap. 2. The UK aid agency, DFID, has been a significant stakeholder in this evolution and given that they tend to focus their activities primarily on Commonwealth countries then this apparent bias in the literature may perhaps be anticipated. What is also interesting from Fig. 5.1 is the relative paucity of reported SLA's in South America. It is also worthy of note that SLAs based in Europe and North America are almost entirely absent. One of the notable examples where SLA was utilised in a developed world country is provided by Davies et al. (2008) in Australia, where it was applied to aboriginal communities. Thus the message from Fig. 5.1 is that SLA has largely been applied in the developing world (the 'Global South') and particularly to countries that were once part of the British Empire.

The mirror of the picture represented in Fig. 5.1 is the country of residence of the authors of the studies, or more accurately the country where the authors' institution is located. These are different given the international nature of research institutes and universities. In the UK for example, a report written by 'Universities UK' (an umbrella group) on the international market for academic staff with the

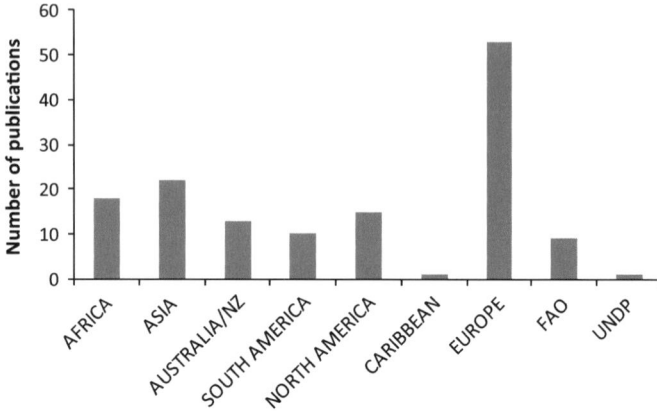

Fig. 5.2 Authorship of SLA publications

evocative title 'Talent Wars' published in July 2007 estimated that in the academic year 2005/2006 some 19 % of academic staff in the UK were non-UK nationals (some 27 % of all academic staff appointed in 2005/2006 were non-UK nationals). This indicates a highly mobile workforce, at least for the UK, and in journal papers the authors provide the address of the employer at the time of the work and not their birthplace. However, an analysis is still useful in providing a sense of the places that 'do' the SLAs and how this might differ from the places where SLAs are 'done'. Figure 5.2 presents counts of 'SLA authors' for the regions where the authors were 'located'. The 70 papers had a total of 142 authors as many had more than one author (not surprising given the complex nature of SLA). European-based authors dominate this list; representing 37 % of the total author count. Indeed of the 53 authors based in Europe some 31 (22 % of the total author count) are from the UK. By way of comparison, authors from the whole of Africa, Asia, South America and the Caribbean combined (representing a substantial part of the Global South), numbered 51 in total (36 % of the total count of 142); a figure equivalent to that of Europe.

Therefore it is not unreasonable to claim that Figs. 5.1 and 5.2 together paint a picture of SLA as a predominantly developed world concept applied to the developing world, and given its origins in 'intentional 'development then this is perhaps to be expected. SLA is therefore, a framework by which outsiders can understand the livelihoods of those which previously, they may have had little, if any, previous contact or knowledge as a first step towards trying to help them. In effect, SLA was created within a context of a rich 'us' trying to help a poor them poor 'them' and this history brings with it some unintentional baggage. In theory SLA should be applicable to all human beings with a livelihood, irrespective of where they live, yet Fig. 5.1 does not suggest that this is the case; SLA is apparently not for the developed world. Why is that so? This will be discussed later, but this history within 'intentional development' could act to restrict its assumed usefulness within the developed world. Thus SLA could to some extent be a prisoner of its own history.

In fairness it has to be stressed that the geographical dichotomy in conceptualisation-implementation observed within the SLA literature is by no means the only or indeed the first such example in the broader development literature. The reader needs to look no further than the evolution of stakeholder 'participation' with its origins in Rapid Rural Appraisal (RRA) from the 1970s, or indeed earlier, as a means by which the poor can be included within analysis and decision-making. Stakeholder participation did not begin with RRA in much the same way that notions of capital and resilience did not begin with SLA, but it represented a new wave within intentional development. Partly this was because of a predominant 'top down' approach to development that existed till the advent of RRA and partly because RRA used new visual tools designed to engage the poorest and less educated in society. Indeed SLA in practice often utilises the sort of tools used within RRA such as mapping and group discussions, and these were employed in the case study set out in this book.

5.4 Transferability of SLA

As discussed in Chap. 2 SLA does encourage the following:

- a wide perspective
- an engagement with people
- an understanding of current livelihood as well as opportunities and threats
- an appreciation of change over time and why that has happened
- an appreciation of what an intervention should be and how it would help to improve matters

These are all appealing attributes. The taking of a wide perspective and an engagement with those meant to benefit makes sense, as does the need to understand how livelihoods are currently constructed and the options people have for improvement. Indeed such an appeal should make SLA applicable and desirable in any context but as noted above this does not seem to have been the case. The reasons are probably varied. To begin with there is often a 'glass fence' divide between the mandate of national and international-focussed agencies in many countries. For example, in the UK, DFID has a mandate for international development that includes environment and agriculture, while the Department for the Environment, Food and Rural Affairs (DEFRA) currently has the mandate for these (fields) amongst others in the UK. Approaches and techniques developed and/or adopted by DFID for its work may not necessarily find their way into DEFRA and vice versa. But the 'glass fences' do not stop there. Higher education institutions and research institutes in the UK and elsewhere often have similar internal arrangements with different departments involved in projects from the developed and developing worlds. These groups may even publish in different journals, and the resulting literatures may not necessarily be familiar to academics and researchers from both sides of the 'glass fence'. Therefore it is perhaps not unreasonable to suppose that as SLA originated from one side of the fence then it may not necessarily be familiar to those on the other side.

The glass fence argument applied to institutions may be one possible explanation for the geographical divide seen with SLA but this is highly simplistic and cannot be the sole explanation. There may also be the attraction of an approach which can be applied by outside experts to a community to help them understand and help that community. 'Outside' in this context does not only mean the geography of where they live but also their culture, language, wealth and so on. Thus there is a sense of the 'alien' that needs to be studied and understood and SLA neatly fits that demand. For practitioners that are trying to bring about change within a context they believe they understand—in their own cultural context for example—then the need for something as intensive, intrusive and demanding as SLA may perhaps be far less obvious and the response may be to turn to other approaches which they think to be more appropriate. As is so often the case in research if the starting point is a set of 'knowns' (or assumed 'knowns' = hypotheses or expectations) that cover aspects of livelihood then it may only be necessary to fill in some gaps rather than follow the whole process. In effect they may see SLA as being a very large hammer to crack a small nut. In practice, of course, this may be debateable as even within a single country livelihoods may not be as well understood as researchers may think. SLA may be complex and resource-intensive but the framework is designed to encourage an exploration of the unexpected and not just the unknown.

Thirdly there could be an element of overlap with the argument often put forward for the rise of participatory techniques in general. These techniques have often been portrayed as helping to provide a 'voice' for communities in situations where democracy is weak, or non-existent, with powerful elites able to impose change at will. Here policies can be created without any involvement of those potentially impacted upon, and this includes the setting of research agenda and creation of new technologies. At least in theory the use of SLA can help create a bridge between local communities, help in assessing what they need, along with evidence of that need and gradually presenting this chain of reason to those with the power to bring about change. In the developed world with more sophisticated and presumably effective democratic structures that link communities to decision makers, then the rationale for approaches such as SLA may be less obvious for if people don't like their leaders they can vote them out of office.

Fourthly, and arguably perhaps the weakest point, nonetheless worth mentioning, livelihoods in the developed world may by and large be simpler than those of the developing world. After all, for the majority of people in the richer parts of the world, livelihood comes down to a wage or salary. Sure there are complications such as concerns about the security of one's job, and how that may be influenced by the wider economic situation, but the heart of livelihood may not be as multifaceted as in poorer regions. But in many countries the aspects of such simpler livelihoods are well explored and appreciated. As Tao and Wall

(2009) have pointed out SLA does seem to work best where there are multiple contributions to livelihood rather than just a single wage or salary. Interestingly, in the latter case SLA may even be seen by researchers as over-elaborate and unnecessary. This view is contentious, of course, as while the source of a livelihood may well be narrower in some places than in others there is still a need to consider resilience and what can enhance or limit it. A narrow source of livelihood could arguably increase vulnerability.

While the above points may not necessarily be the only ones at play in helping to create this geographical divide in the use of SLA, there are other caveats that need to be considered. Most notably just because the SLA framework as formally set out by DFID and others has not been reported as such for the developed world, that is not the same as saying that the underlying ideas it encompasses have not been applied. The point has already been made in the context of DDS that over its history it may not have used the formal terminology associated with SLA but it had intuitively applied many of its principles. Similarly, the notions of 'capitals' at the heart of the SLA are by no means unique to that framework and have often been explored in the developed world. Similarly the idea of exploring the role of institutions in peoples' lives is certainly not unique to the developing world and neither is the notion of livelihood resilience, even if these are often expressed in diverse ways. Based on the DDS experience, it is important to understand that while the SLA may be presented as a package, its components and interactions are often part of many other analyses. The expression of an SLA framework may be geographically biased but the ideas at its heart are not.

5.5 Livelihood into Lifestyle

The reader may well have picked up from the preceding pages of this book the challenges of SLA in practice. The breadth of the approach is logical, as people can have livelihoods that encompass many interacting capitals all embedded within a rich context of history, institutions and so on. As Sanderson (2009) has put it in response to those who are critical of livelihood approaches such as SLA:

> Those that criticise livelihoods approaches point to their breadth—livelihoods can be almost anything. But that is precisely the point. SLA provides a model for navigating messy reality, and for layering onto that the range of development and emergency interventions, always with people at the centre. That in itself is more than enough for any piece of thinking.

This is very true; the breadth of SLA is certainly one its strengths and should allow it to be applicable in just about any development context, or indeed any context where it is necessary to understand livelihood. But in practice it can also be a weakness. A framework that can help see the holism (or *messy reality*) of what is involved with livelihood is useful as a conceptual device but it doesn't wash away the complexity when it comes to the practicalities of putting the framework into practice. This is not to say that SLA is wrong or should be rejected but only that it is difficult and the effort involved should not be underestimated. The DDS experience with SLA was at two levels. Firstly it can be argued that DDS had employed SLA (seen as a set of principles) throughout much of its existence, even if it wasn't given that name. Secondly the use of SLA (as a framework) for the two villages each took a year to complete with all the challenges outlined in Chap. 4. In both cases—framework for analysis and a set of principles—time was on the side of DDS. The reader may feel that seeing SLA as a set of principles is vague, and indeed this is true given how broad and logical those principles are, but he application of SLA to the two villages in the sense of it being a practical framework for guidance was also time and resource consuming. There are dangers with the implementation of an SLA as already highlighted in the trade-offs faced by DDS. But again that is not to say that the framework is wrong but simply that it is difficult to implement and there are dangers inherent in short cuts and especially incorrect interpretations of information that may result from poor SLA practices. This is true of much research in the social and economic sciences, and the point has already been made in Chap. 2 regarding the constructivist/interpretivist position where it is argued by some that there are real dangers is treating it in a way which suggests that it can be deconstructed so as to arrive at cause-effect relationships. While the points made about such dangers are noted the alternative may be to do nothing— to make no attempt to understand livelihood and what can be done to help make them sustainable. Mistakes are possible given the complexity involved but having a framework for guidance is more desirable than having no help. Criticising the breadth of SLA and highlighting the pitfalls should not be construed as meaning that nothing should be done and that livelihoods should remain unexplored.

The authors wish to go further and show that SLA for all its breadth does not arguably go far enough and misses much of importance in sustainability. Carney's (1998) definition of livelihood has already been given in Chap. 2 but given its content it is worth repeating:

> A livelihood comprises the capabilities, assets (including both material and social resources) and activities required for a means of living.

The key phrase here is *means of living*. But the point has already been made that one of the limitations of SLA is its inherent focus on livelihood to the exclusion of other important aspects of human existence. People don't just 'do' livelihoods—they don't just have a *means of living*—but there are other considerations such as culture, recreation, family, social status and so on which come into play each and every day and which can impact upon livelihood. Indeed it can be said that livelihoods are influenced by values. Books and papers on sustainability normally begin by providing a formal definition as to what it is and while there are many of them the one most used is that of the World Commission for Environment and Development. Here it is given at the end of the book rather than at the start and there is a reason for that. The WCED definition is as follows:

> development that meets the needs of the present without compromising the ability of future generations to meet their own needs.
> WCED (1987), p. 8

Note how this definition ends in the plural *needs*; in fact the word is used twice in the definition and linked to current and future generations. This embodies something much wider and more comprehensive (holistic) that just a *means of living*. People's lives are not static no matter how poor; there is far more to life however poor one is than mere survival. People have dreams and expectations and seek ways as to how their livelihood can be enhanced to meet their aspirations. Thus any analysis of livelihood has to take cognisance of these dreams and ambitions as these help frame where people wish to be. There has to be a sense of progress within any appreciation of livelihood, and the process should encourage such exploration of where people want to be and indeed influence it. A pitfall here is that in helping someone meet such ambitions, the result may be detrimental to others, including future generations.

Some other components such as recreation and culture may not necessarily help support livelihood in a direct sense, but are nonetheless important. In short, people have a lifestyle as well as a livelihood, where lifestyle can be defined in simple terms as the ways in which a person or group lives. Livelihood is a means by which people underpin their lifestyle, but the two are obviously related. Included here, are the accumulations of goods that people may not need for 'livelihood' but which they perceive as enhancing their status in society—or in effect how others see them. The SLA framework as set out in Fig. 2.1 does not in itself embrace these wider concerns of lifestyle, although there may be impacts upon one or more capitals and much (admittedly) depends upon definition. It is not difficult to see how these become intertwined. For example, an individual's involvement in a club or society for recreational purposes could bring them into contact with others who could help them in terms of livelihood. A number of societies mentioned under 'social capital' for the two villages in Chap. 4 are primarily recreational. But such a gain for livelihood from recreation may not always be possible, and indeed within an SLA such activities may not be regarded as relevant.

Prosperity is often defined quite narrowly in terms of income but could equally be regarded as having the lifestyle one desires. This is more than a material concern of owning enough goods. As Tim Jackson has pointed out:

> Prosperity is about things going well for us—in accordance with (pro- in the Latin) our hopes and expectations (speres). Wanting things to go well is a common human concern. It's understood that this sense of things going well includes some notion of continuity. We are not inclined to think that life is going well, if we confidently expect things to fall apart tomorrow. There is a natural tendency to be at least partly concerned about the future. Jackson (2009), p. 16

Indeed the research agenda has arguably moved more into understanding and enhancing the sustainability of lifestyle—making sure those things don't fall apart tomorrow—rather than a focus on livelihood. Part of this is about understanding the wants of lifestyles and how that impacts upon the resources available on planet earth and other people and species that inhabit this same space. Thus lifestyle is more in tune with concerns over consumption, whereas livelihood has concerns about production and consumption. Indeed the notion of livelihood is arguably more in tune with the emphasis one sees so often at national and regional scales where a maximisation of indicators such as the Gross Domestic Product (GDP) is king and the tendency of governments has been to promote consumerism; albeit to different degrees in different countries. The logic is a clear one; emphasise income generation as a means of underpinning livelihood and then all else will follow. Others, such as Tim Jackson quoted above, have argued against this dominant focus on increasing economic growth, at least for the developed world. But can the developed world accept this, especially as monetary income is still seen by many (including governments) as the linchpin for the lifestyle?

A question which follows on from the above is whether the 'classic' SLA framework should be modified to include a sense of where people want to be with their livelihood and how this is driven by the lifestyle they wish to enjoy? These considerations can certainly be added to the 'classic' SLA framework as shown in Fig. 5.3 where lifestyle, and all its associations, has been highlighted (not implied) as both a key driver of livelihood and underpinned by it. There is the inclusion of lifestyle expectations (what people would like) as well as their current lifestyle), and interventions that are intended to change some of these if necessary. An obvious dilemma with Fig. 5.3, herein referred to as SLifA (Sustainable Lifestyle Analysis) is that while a need to influence lifestyle choices may be desirable (perhaps even imperative) in the richer countries of the world it is less palatable in poorer regions. How can it be desirable to destroy an expectation from a family

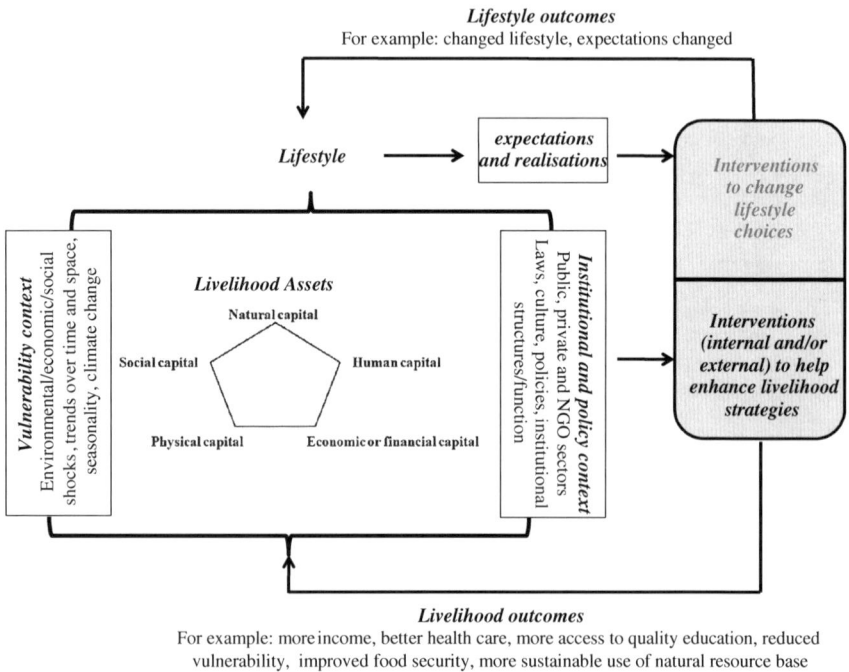

Fig. 5.3 The sustainable lifestyle approach (*SLifA*)

that wish to own a car? How can this be squared with the high level of car ownership in richer countries? In both cases the car may be needed for livelihood (the
example of commuting has already been given) but also for recreation and status.
Indeed while SLifA raises deep issues about growth in the developing world, and
the point has already been made that it is hard to see how lifestyle expectations of
the poorest communities in the world can be limited, it does not necessarily mean
that such limits will be any easier to embrace in the developed world.

As with the classic SLA diagram (Fig. 2.1) Fig. 5.3 is a theoretical framework
which sets out points that need to be considered. It certainly does not make it
easier to put into practice and if anything makes it far more difficult. But these
are important and challenging questions none the less, and SLifA helps to bring
into the classic SLA the important (and missing) domain of ethics. SLifA certainly
makes the process more messy and complex, but even with the issues highlighted
here it is debateable whether that in itself is a bad thing. After all, the logical
endpoint of that argument is for there to be no attempt at all to provide help to
a community. As already stated, complexity should not be an excuse for disen-
gagement. It is certainly the case that over-simplification can be dangerous, and
the DDS example has shown how easy it is to come to superficial but incorrect
data. Each of their SLAs took a year to complete with much triangulation to check
answers researchers received. One of the problems of SLA is that its complexity
can encourage the 'quick fix', especially when time and resources are limited and
that will be no different with SLifA. But what it does do is extend the boundary of
what needs to be considered.

An interesting outcome of SLifA is that it brings a model designed by the
developed for the developing world into a broader realm of application that dis-
misses that polarity. At first this might sound counter intuitive as the SLA
framework should be applicable to people wherever they live, but some of the reasons why the SLA has the geographic dipole it has have already been discussed.
SLifA may not necessarily change that polar reality but it should. Lifestyle and
the consumption that underpins it are concerns for everyone, even if there are
important issues of equity and social justice at play. Indeed the inclusion of lifestyle may resonate far more with developed world contexts, and also opens up the
need to change behaviours which are damaging to society and the environment.
Thus while SLA speaks of interventions in terms of improving the lot of the poor,
SLifA also includes the need to consider whether all behaviours, not just those
underpinning livelihood, need to be changed. The refection-action reflection praxis
is no longer about what actions can be done to improve livelihood but about lifestyle, and some of this could be about what the household itself can do.

It is perhaps worthy of note that DDS pursued a reflective approach
(reflection-action reflection) which was far more akin to the SLiFA model than it
was to SLA; at least when seen as a set of guiding principles. Thus it was not just
a case of understanding livelihood and working to improve it as DDS was also
engaged in other aspects of human existence including the spiritual and appreciating all circumstances within the household especially ambitions, anxieties and
concerns of those that comprised it. In the course of evaluating DDS services

in 1997 the participants rated the extra-curricular activities such as care of the physically and mentally challenged and the HIV-AIDS programmes higher than the micro-finance and agricultural components. DDS was seen as a source of advice, guidance, information and assistance rather than just an agency which helped with improving livelihood. But this took time to develop and SLiFA is perhaps an approach that takes a long -term engagement for it to succeed and may not be suited to short-term and project focussed planning and intervention. There needs to be continuous dialogue and discussion as to what is happening not just in the physical sense but around how this change is affecting thinking and behaviour.

Finally it is necessary to return to a point made towards the end of Chap. 2 regarding the recent retreat of DFID from the use of SLA (Clark and Carney 2009). Of the reasons behind this retreat of one of the prominent pioneers of SLA was its apparent association with the 'project level' and this is, of course, based on a vision of SLA as a practical framework rather than a set of principles. DFID appears to have moved its focus more towards national-scale interventions and SLA was apparently perceived by them as having little to offer at that scale. As a result Clark and Carney (2009) suggest a number of strategies to enhance the relevance of SLA, and at least to the authors of this book these strategies say a lot about the often unwritten weaknesses of SLA. The fact that Clark and Carney felt that these changes were needed, at least in part, to help enhance adoption of SLA within DFID is interesting in itself. But some of their other suggestions also serve to highlight the complexity of 'doing' SLA rather than pointing to a problem with the ideas. Two of them are as follows, along with some notes as to how an incorporate of lifestyle within the traditional SLA could help.

- *review how SLA can be adapted to contribute to current development challenges, including the food crisis, fragile states, economic growth and making markets work for the poor*

It has already been noted how SLA should be able to readily encompass these challenges and indeed any that can be imagined within development. As its champions have stressed, the holistic nature of SLA should allow it to cope with any 'messy reality', and indeed the SLA literature is replete with its application in a wide variety of contexts. Hence it is intriguing as to why Clark and Carney feel that changes are required to SLA to allow it to better contribute to these challenges—as complex and intractable as they are. Perhaps this is a further manifestation of the 'scale' issue—that SLA is too micro in its spatial focus—but why should this inherently be so? Surely an SLA approach can be used to explore actors within a food supply chain or markets? The people in those systems might be separated in space but they have livelihoods and presumably a desire a make them more sustainable.

An adaptation of livelihood that includes a consideration of lifestyle and influences (including policy) that may be required to help manage expectations within lifestyle could be one means of achieving such a wider relevance. Some points surrounding 'theory of change' and 'evidence-based' approaches have already been made in Chap. 2, and influences move in a number of directions across actor groups.

- *address perceived weaknesses of SLA, such as limited analysis of policy processes, ecological sustainability, gender and power relations.*

Parts of the latter point may seem odd as SLA in its classic form and as applied in Chap. 4 should include *ecological sustainability* (overlap here with ecosystems goods and services discussed in Chap. 2) within natural capital and elsewhere, and similarly *gender and power relations* should also appear when it comes to social and human capital as well as wider institutional and resilience contexts. It is hard to see how these are *perceived weaknesses* in SLA. One wonders whether the issue here is more the complexity of addressing these within SLA rather than the idea that they should be. After all, as the Nigerian case study has shown it can be difficult and time-consuming to get at the bottom of some of these. The importance of ethnicity for land ownership was shown for Ekwuloko, and this was not something that anyone attempted to hide or minimise, but power relations can be difficult to fathom. But a broadening of SLA to SLiFA would help highlight an important role for policy makers and politicians, and should reinforce the need for a consideration of ecological sustainability (lifestyle has a strong link to consumption) as well as gender and power relations.

5.6 Conclusions

This book is about SLA and how it was conceived and applied which is in itself a rather narrow outlook. It has covered the evolution and application of an approach that is intended to lead to progress; a new way of thinking about livelihood as well as a new way of intervening to help improve livelihood and enhancing sustainability. But the story has encompassed a number of issues of import, indeed central, in the sustainable development discourse, from an emphasis on livelihood rather than lifestyle to a geographical dichotomy of those having the power to set the SLA agenda and those who are less powerful and who are the focus of the analysis. SLA is a microcosm of this.

While SLA has been popular amongst development agencies and researchers it may be that its popularity is waning as evidenced by the DFID retreat—one of its original and key advocates. The number of publications mentioning SLA has not been high—70 in total from 1999 to mid 2012 (13 years) is not a large output although it is also not insignificant—and there is no evidence that the publication-rate is increasing. For a holistic approach that was so promising the last mentioned facts are hardly encouraging. Could it be that the future of SLA is uncertain and it is certainly not difficult to understand why this might be so. Some have pointed to the lack of applicability of SLA as a practical framework to larger-scales such as the nation-state as one reason, but there are no doubt other concerns related to the observed polarity of SLA as an approach promoted by the Global North for the Global South. SLA should be applicable in any situation where livelihood needs to be understood and enhanced so as to make it more sustainable, but in practice its use has been restricted to the developing world.

In addition to this is the practical challenge of 'doing' an SLA as highlighted with regard to the case study in Nigeria, and the tensions between cost, quality and representation have been mentioned. The way DDS did its SLA may have suited its purposes and circumstances but is by no means the only way in which an SLA can be employed. There are indeed cheaper and quicker ways of collecting data to populate the SLA framework but issues of data quality may then begin to emerge. But this is probably only a small part of the story. Much of the results of SLAs may not appear in journal publications but as reports and these may not necessarily be accessible to those outside the agencies which commissioned them. But whichever way one looks at it, the evidence suggests that the future of 'classic' SLA as set out in Chap. 2 may not be a rosy one. This is ironical given the logic setting within which the SLA is embedded, but it may also be the case that the agenda is moving on—or trying to. There is an increasing questioning of the desirability let alone practicality of economic growth based upon ever increasing consumption, especially as governments continue to prioritise it. It now appears the critical question now revolves more around how lifestyle is managed rather than livelihood. Prosperity as part of lifestyle and how to maintain it may be replacing an emphasis on income and goods; livelihood is after all only one part of the human experience.

However, many people still face poverty and for them an enhancement of livelihood so as to make it both more rewarding and more sustainable is a priority. SLA was created to help these people within the context of an intentional approach to development. The logic upon which it is founded is credible, even if its practice may be imperfect; it is hard to argue against the principles which are encompassed within the framework. Indeed the point has been made that field agencies working at the cutting edge of development have long incorporated the principles set out in the SLA. What is perhaps needed is an extension of the framework to make the links with policy more apparent and the SLiFA framework which includes a consideration of lifestyle; management of expectations is perhaps one way forward. There is nothing new here but such management impels leaders and the better off members of the community to lead by example especially with respect to the 'consumption' side of the equation. Personal responsibility has to be challenged which necessitates drawing upon important issues such as ethics, distribution and social justice. In other words SLiFA not only embodies SLA but a whole range of spiritual and moral obligations although in so doing it does not make development easier. SLiFA adds many more considerations to the list provided within SLA, and that is the challenge. On the other hand, human beings resist simplification so SLA has helped blaze an important trail by guiding the discourse to this point demonstrating as it did so that there are no simple fixes in any intervention.

Finally, it perhaps has to be stressed that the power of frameworks such as SLA and indeed SLiFA is not so much to provide mechanistic and almost 'tick box' recipes for investigation and intervention but as concepts that help guide awareness. The relatively small number of papers published on SLA mentioned above is mostly those that refer to SLA more in its 'recipe' mode for a specific project or programmes. The wider concept of a 'sustainable livelihood' has been much more popular and the same is true of a 'sustainable lifestyle'. SLA is to the former what SLiFA is to the latter. These frameworks can be seen as enhanced definitions rather than recipes and in that sense have been successful even if they have not been noted; they can be present without being visible. That is perhaps their greatest contribution.

References

Ahmed, N., Troell, M., Allison, E. H., & Muir, J. F. (2010). Prawn post larvae fishing in coastal Bangladesh: Challenges for sustainable livelihoods. *Marine Policy, 34*(2), 218–227.

Ahmed, F., Siwar, C., & Idris, N. A. H. (2011). The sustainable livelihood approach: reduce poverty and vulnerability. *Journal of Applied Sciences Research, 7*(6), 810–813.

Akinkugbe, F. (1976). An internal classification of Yoruboid Group (Yorùbá Itsekiri, Igala). Journal of West African Languages, *xi*, 1–2

Allison, E. H., & Ellis, F. (2001). The livelihoods approach and management of small-scale fisheries. *Marine Policy, 25*(5), 377–388.

Alison, E. H., & Horemans, B. (2006). Putting the principles of the Sustainable Livelihoods Approach into fisheries development policy and practice. *Marine Policy, 30*(6), 757–766.

Amalric, F. (1998). *The Sustainable Livelihoods Approach: general report of the sustainable livelihoods project 1995–1997*. Rome: Society for International Development.

Armstrong, D. (2006). *Organisation in the mind: psychoanalysis, group relations, and organisational consultancy*. London: Karnac Books.

Ashley, C., & Carney, D. (1999). *Sustainable Livelihoods: lessons from early experience*. London: Department for International Development.

Ashley, C., Chaumba, J., Cousins, B., Lahiff, E., Matsimbe, Z., Mehta, L., Mokgope, K., Mombeshora, S., Mtsi, S., Nhantumbo, I., Nicol, A., Norfolk, S., Ntshona, Z., Pereira, J., Scoones, I., Seshia, S., Wolmer, W., & Nyamu-Musembi, C. (2003). 8. Rights talk and rights practice: Challenges for southern Africa. IDS Bulletin—Institute of Development Studies 34(3), 97–111

Banister, D., & Stead, D. (2004). Visioning and back casting: desirable futures and key decisions. Paper presented at the STELLA FG4 meeting, Brussels, Belgium, 25–27 March 2004.

Bebbington, A. (1999). Capitals and capabilities: A framework for analyzing peasant viability, rural livelihoods and poverty. *World Development, 27*(12), 2021–2044.

Bebbington, A. (2002). Social Capital/Social Development/SDV. Note prepared for the workshop: Social Capital: The value of the concept and strategic directions for World Bank Lending, IFC, Washington DC, 1 March 2002. Washington DC: World Bank.

Bell, S., & Morse, S. (2008). Sustainability Indicators. Measuring the immeasurable (2nd ed.). London: Earthscan.

Below, T. B., Mutabazi, K. D., Kirschke, D., Franke, C., Sieber, S., Siebert, R., et al. (2012). Can farmer's adaptation to climate change be explained by socio-economic household-level variables? *Global Environmental Change—Human and Policy Dimensions, 22*(1), 223–235.

Berry, S. (1989). Social institutions and access to resources. *Africa, 59*(1), 41–55.

Black, N. (2001). Evidence based policy: Proceed with care. *British Medical Journal, 323*, 275–278.

Blaikie, P. (1989). Environment and access to resources in Africa. *Africa, 59*(1), 18–40.

S. Morse and N. McNamara, *Sustainable Livelihood Approach*,
DOI: 10.1007/978-94-007-6268-8,
© Springer Science+Business Media Dordrecht 2013

Blaikie, P. (2000). Development, post-, anti-, and populist: A critical review. *Environmental and Planning A, 32*, 1033–1050.

Bondad-Reantaso, M. G., Bueno, P. B., Demaine, H., & Pongthanapanich, T. (2009). Development of an indicator system for measuring the contribution of small-scale aquaculture to sustainable rural development. *FAO Fisheries and Aquaculture Technical Paper, 534*, 161–179.

Boston, J. S. (1968). *The Igala Kindom*. Nigeria: Oxford University Press for Nigerian Institute of Social and Economic Research.

Boswell, C. (2008). The political functions of expert knowledge: Knowledge and legitimation in European Union immigration policy. *Journal of European Public Policy, 14*, 471–488.

Boyd, J. W., & Banzhaf, H. S. (2005). Ecosystem services and government: The need for a new way of judging nature's value. *Resources, 158*, 16–19.

Brent, A. C., & Kruger, W. J. L. (2009). Systems analyses and the sustainable transfer of renewable energy technologies: A focus on remote areas of Africa. *Renewable Energy, 34*(7), 1774–1781.

Brinson, A. A., Die, D. J., Bannerman, P. O., & Diatta, Y. (2009). Socioeconomic performance of West African fleets that target Atlantic billfish. *Fisheries Research, 99*(1), 55–62.

Bruckmeier, K., & Tovey, H. (2008). Knowledge in sustainable rural development: From forms of knowledge to knowledge processes. *Sociologia Ruralis, 48*(3), 313–329.

Bueno, P. B. (2009). Indicators of sustainable small-scale aquaculture development. *FAO Fisheries and Aquaculture Technical Paper, 534*, 145–160.

Butler, L. M., & Mazur, R. E. (2007). Principles and processes for enhancing sustainable rural livelihoods: Collaborative learning in Uganda. *International Journal of Sustainable Development and World Ecology, 14*(6), 604–617.

Campbell, D. T. (1975). Degrees of freedom and the case study. *Comparative Political Studies, 8*(1), 178–191.

Campbell, D. T., & Stanley, J. C. (1966). *Experimental and quasi-experimental designs for research*. Chicago: Rand McNally.

Carney D (Ed.) (1998). Sustainable rural livelihoods. What contribution can we make? Department of International Development. Nottingham: Russell Press Ltd.

Carney, D., Drinkwater, M., Rusinow, T., Neefjes, K., Wanmali, S., & Singh, N. (1999). *Livelihoods approaches compared*. London: Department for International Development.

Carpenter, S. R., DeFries, R., Dietz, T., Mooney, H. A., Polasky, S., Reid, W. V., et al. (2006). Millennium ecosystem assessment: Research needs. *Science, 314*, 257–258.

Carswell, G. (1997). Agricultural intensification and Sustainable Rural Livelihoods: A think piece. IDS Working Paper 64. Brighton: Institute for Development Studies.

Challies, E. R. T., & Murray, W. E. (2011). The interaction of global value chains and rural livelihoods: The case of smallholder Raspberry growers in Chile. *Journal of Agrarian Change, 11*(1), 29–59.

Chambers, R. (1989). Editorial introduction: Vulnerability, coping and policy. *IDS Bulletin, 20*(2), 1–7.

Chambers, R. (1991). *Rural development: Putting the last first*. London: Intermediate Technology.

Chambers, R., & Conway, GR. (1992). Sustainable rural livelihoods: Practical concepts for the 21st century. IDS Discussion Paper No. 296. Brighton: IDS.

Chang, Y., & Tipple, G. (2009). Realities of life and housing in a poor neighbourhood in urban China Livelihoods and vulnerabilities in Shanghai Lane, Wuhan. *International Development Planning Review, 31*(2), 165–198.

Cherni, J.A., Dyner, I., Henao, F., Jaramillo, P., Smith, R., & Font, R.O. (2007). Energy supply for sustainable rural livelihoods. A multi-criteria decision-support system. Energy Policy 35(3), 1493–1504.

Cherni, J. A., & Hill, Y. (2009). Energy and policy providing for sustainable rural livelihoods in remote locations: The case of Cuba. *Geoforum, 40*(4), 645–654.

Choi, B. C. K., Pang, T., Lin, V., Puska, P., Sherman, G., Goddard, M., et al. (2005). Can scientists and policy makers work together? *Journal of Epidemiology and Community Health, 59*, 632–637.

Clark, J., & Carney, D. (2009). Revitalising the sustainable livelihoods approach. DFID: London. Available at www.eldis.org/id21ext/SLA/ClarkCarneyFeb09.html

Conroy, C., & Litvinoff, M. (Eds.). (1988). *The greening of Aid: Sustainable Livelihoods in practice*. London: Earthscan Publications Ltd.

Conway, G. R. (1985). Agro ecosystem analysis. *Agricultural Administration, 20*, 31–55.

Costanza, R., d'Arge, R., de Groot, R., Farberk, S., Grasso, M., Hannon, B., et al. (1997). The value of the world's ecosystem services and natural capital. *Nature, 387*, 253–260.

Cowen, M., & Shenton, B. (1998). Agrarian doctrines of development: Part 1. *Journal of Peasant Studies, 25*, 49–76.

Crouch, M., & McKenzie, H. (2006). The logic of small samples in interview-based qualitative research. *Social Science Information, 45*(4), 483–499.

Daily, G. C. (1997). Introduction: What are ecosystem services? In G. C. Daily (Ed.), *Natures services: Societal dependence on natural ecosystems* (pp. 1–10). Washington, D.C.: Island Press.

Daramola, A., Ehui, S., Ukeje, E., & McIntire, J. (2008). Agricultural export potential in Nigeria. In Paul Collier, Chukwuma C Soludo & Catherine Pattillo (Eds.), *Economic policy options for a prosperous Nigeria*. Palgrave: MacMillan. Available at www.csae.ox.ac.uk/books/epopn/AgriculturalexportpotentialinNigeria.pdf.

Daskon, C., & Binns, T. (2010). Culture, tradition and sustainable rural livelihoods: Exploring the culture-development interface in Kandy, Sri Lanka. *Community Development Journal, 45*(4), 494–517.

Davies, S. (1996). *Adaptable livelihoods: Coping with food insecurity in the Malian Sahel*. London: Macmillan.

Davies, J., White, J., Wright, A., Maru, Y., & LaFlamme, M. (2008). Applying the sustainable livelihoods approach in Australian desert Aboriginal development. *Rangeland Journal, 30*(1), 55–65.

Dawtry, B. (Ed.) (1980). Elaeis. A Journal of Igalaland, 2(2).

de Groot, R. S., Wilson, M. A., & Boumans, R. M. J. (2002). A typology for the classification, description and valuation of ecosystem functions, goods and services. *Ecological Economics, 41*(3), 393–408.

de Groot, R., Fisher, B., Christie, M., Aronson, J., Braat, L., Gowdy, J., et al. (2010). Integrating the ecological and economic dimensions in biodiversity and ecosystem service valuation. In P. Kumar (Ed.), *The economics of ecosystems and biodiversity: Ecological and economic foundations*. Oxford: Earthscan.

de Haan, LJ. (2005). How to research the changing outlines of African livelihoods. Paper presented at the 11th General Assembly of CODESRIA, 6–10 Dec 2005, Maputo 2005.

Department for Environment Food and Rural Affairs. (2007). *An introductory guide to valuing ecosystem services*. London: Department for Environment, Food and Rural Affairs.

Department for International Development (DFID) (1997). Eliminating World Poverty: A Challenge for the 21st Century. White Paper on International Development. London: HMSO. Available at www.dfid.gov.uk/Pubs/files/whitepaper1997.pdf

Department for International Development (DFID). (1999). *Sustainable livelihoods and poverty elimination*. London: Department for International Development.

Department for International Development (DFID). (2000a). *Sustainable livelihoods: Current thinking and practice*. London: Department for International Development.

Department for International Development (DFID). (2000b). *Sustainable livelihoods: Building on strengths*. London: Department for International Development.

Department for International Development (DFID) (2000c). Achieving sustainability: Poverty elimination and the environment, strategies for achieving the international development targets. London: Department for International Development.

Department for International Development DFID. (2000). *Eliminating World Poverty: Making globalisation work for the poor, white paper on international development*. London: Stationery Office.

D'Silva, B. C., & Raza, M. R. (1980). Integrated rural development in Nigeria: The Funtua project. *Food Policy, 5*(4), 282–297.

Diamond, P. A., & Hausman, J. A. (1994). Contingent valuation: Is some number better than no number? *The Journal of Economic Perspectives, 8*(4), 45–64.

Dogan, M., & Pelassy, D. (1990). *How to compare nations: Strategies in comparative politics* (2nd ed.). Chatham UK: Chatham House.

Eboh, E. C. (1995). Informal financial groups in Nigeria: Organization, management and the implications for formal finance interventions. *Journal of Rural Development, 14*(2), 137–149.

Edelman, G. M. (2003). Naturalizing consciousness: A theoretical framework. *Proceedings of the National Academy of Sciences of the United States of America, 100*(9), 5520–5524.

Ehrlich, P., & Ehrlich, A. (1981). *Extinction: The causes and consequences of the disappearance of species.* New York: Random House.

Elasha, B.O., Elhassan, N.G., Ahmed, H., & Zakieldin, S. (2005). Sustainable livelihood approach for assessing community resilience to climate change: Case studies from Sudan. Assessments of Impacts and Adaptations to Climate Change (AIACC) Working Paper No. 17

Ellis, F. (2000). *Mixing It: Rural livelihoods and diversity in developing countries.* Oxford: Oxford University Press.

Engel, S., Pagiola, S., & Wunder, S. (2008). Designing payments for environmental services in theory and practice: An overview of the issues. *Ecological Economics, 65*, 663–674.

Erenstein, O. (2011). Livelihood assets as a multidimensional inverse proxy for poverty: A district-level analysis of the Indian Indo-Gangetic Plains. *Journal of Human Development and Capabilities, 12*(2), 283–302.

Escobar, A. (1992). Reflections on 'development': Grassroots approaches and alternative politics in the Third World. *Futures, 24*(5), 411–436.

Escobar, A. (1995). Encountering development: The making and unmaking of the Third World. Princeton: Princeton University Press.

Estreva, C. (1992). Development. In W. Sahs (Ed.), *The development dictionary: A guide to Knowledge and Power* (pp. 6–25). New York: Zed Books.

Falola, T., & Heaton, M. M. (2008). *A history of Nigeria.* Cambridge UK: Cambridge University Press.

Farber, S. C., Costanza, R., & Wilson, M. A. (2002). Economic and ecological concepts for valuing ecosystem services. *Ecological Economics, 41*, 375–392.

Farrington, J. (2001). Sustainable livelihoods, rights and the new architecture of aid. Natural Resource Perspectives 69. London: Overseas Development Institute.

Farrington, J., Carney, D., Ashley, C., & Turton, C. (1999). Sustainable livelihoods in practice: Early application of concepts in rural areas. Natural Resources Perspectives 42. London: Overseas Development Institute.

Fenichel, A., & Smith, B. (1992). A successful failure: Integrated rural-development in Zambia. *World Development, 20*(9), 1313–1323.

Fernandez, L. A. P., Toral, J. N., Vazquez, M. R. P., Barrios, L. G., Beutelspacher, A. N., & Baltazar, E. B. (2010). Impact of income strategies on food security in rural Mayan communities in northern Campeche. *Archivos Latinoamericanos de Nutricion, 60*(1), 48–55.

Fisher, B., & Turner, K. (2008). Ecosystem services: Classification for valuation. *Biological Conservation, 141*, 1167–1169.

Fisher, B., Turner, K., Zylstra, M., Brouwer, R., De Groot, R., Farber, S., et al. (2008). Ecosystem services and economic theory: Integration for policy-relevant research. *Ecological Applications, 18*(8), 2050–2067.

Fisher, B., Turner, R. K., & Morling, P. (2009). Defining and classifying ecosystem services for decision making. *Ecological Economics, 68*(3), 643–653.

Flora, C. B. (1992). Building sustainable agriculture: A new application of Farming Systems Research and Extension. *Journal of Sustainable Agriculture, 2*(3), 37–49.

Flyvbjerg, B. (2006). Five misunderstandings about case study research. *Qualitative Inquiry, 12*(2), 219–245.

Forsyth, D. R. (2006). *Group dynamics.* Belmont, CA: Thomson Wadsworth Publishing.

Freire, P. (1970). Pedagogy of the Oppressed. New York: Continuum

Funnell, S.C., & Rogers, P.J. (2011). Purposeful program theory. Effective use of theories of change. San Francisco: John Wiley and Sons Inc.

Geertz, C. (1995). After the fact: Two countries, four decades, one anthropologist. Cambridge MA: Harvard University Press.

Glavovic, B. C. (2006a). Coastal sustainability: An elusive pursuit? Reflections on South Africa's coastal policy experience. *Coastal Management, 34*(1), 111–132.

Glavovic, B. C. (2006b). The evolution of coastal management in South Africa: Why blood is thicker than water. *Ocean and Coastal Management, 49*(12), 889–904.

Glavovic, B. C., & Boonzaier, S. (2007). Confronting coastal poverty: Building sustainable coastal livelihoods in South Africa. *Ocean and Coastal Management, 50*(1–2), 1–23.

Gaiha, R., Imai, K., & Kaushik, P. D. (2001). On the targeting and cost-effectiveness of anti-poverty programmes in rural India. *Development and Change, 32*(2), 309–342.

Grant, E. (2001). Social capital and community strategies: Neighbourhood development in Guatemala City. *Development and Change, 32*(5), 975–997.

Gray, A. (2001). Evidence-based policies and indicator systems: From profane arithmetic and a Sacred Geometry. Third International Inter-disciplinary Evidence-Based Policies and Indicator Systems Conference, July 2001.

Guyer, J. I. (1981). Household and community in African studies. *African Studies Review, 24*(2/3), 87–137.

Guyer, J. I. (1992). Small change: Individual farm work and collective life in a western Nigerian Savanna town, 1969–1988. *Africa, 62*(4), 465–488.

Guyer, J. I. (1996). Diversity at different levels: Farm and community in West Africa. *Africa, 66*(1), 71–89.

Guyer, J. I. (1997). *An African niche economy*. Edinlburgh: Edinburgh University Press.

Guyer, J., & Peters, P. (1987). Introduction. Conceptualizing the household: Issues of theory and policy in Africa. *Development and Change Special Issue* 18(2), 197–214.

Hanifan, L. J. (1916). The rural school community center. *Annals of the American Academy of Political and Social Science, 67*, 130–138.

Hart, G. (2001). Development critiques in the 1990s: Culs de sac and promising paths. *Progress in Human Geography, 25*(4), 649–658.

Helmore, K., & Singh, N. (2001). *Sustainable Livelihoods: Building on the wealth of the Poor*. Connecticut, USA: Kumarian Press.

Himanen, V., Lee-Gosselin, M., & Perrels, A. (2004). Impacts of transport on sustainability: Towards an integrated transatlantic evidence base. *Transport Reviews, 24*(6), 691–705.

Hogh-Jensen, H., Egelyng, H., & Oelofse, M. (2009). Research in Sub-Saharan African food systems must address post-sustainability challenges and increase developmental returns. *Scientific Research and Essays, 4*(7), 647–651.

Hogh-Jensen, H., Oelofse, M., & Egelyng, H. (2010). New challenges in underprivileged regions call for people-centered research for development. *Society and Natural Resources, 23*(9), 908–915.

Holt, T. (2008). Official statistics, public policy and public trust. *Journal of the Royal Statistical Society Series A: Statistics in Society* 171(2), 323–346.

Huston, A. C. (2008). From research to policy and back. *Child Development, 79*(1), 1–12.

Innvaer, S., Vist, G., Trommald, M., & Oxman, A. (2002). Health policy-maker's perceptions of their use of evidence: A systematic review. *Journal of Health Services Research Policy, 7*(4), 239–244h.

Ison, R. L., Maiteny, P. T., & Carr, S. (1997). Systems methodologies for sustainable natural resources research and development. *Agricultural Systems, 55*(2), 257–272.

Iwasaki, S., Razafindrabe, B. H. N., & Shaw, R. (2009). Fishery livelihoods and adaptation to climate change: a case study of Chilika lagoon, India. *Mitigation and Adaptation Strategies for Global Change, 14*(4), 339–355.

Jackson, T. (2009). *Prosperity without growth? The transition to a sustainable economy*. London, UK: Sustainable Development Commission. Available at www.sd-commission.org.uk/publications.php?id=914

Kelman, I., & Mather, T. A. (2008). Living with volcanoes: The sustainable livelihoods approach for volcano-related opportunities. *Journal of Volcanology and Geothermal Research, 172*(3–4), 189–198.

Knutsson, P. (2006). The Sustainable Livelihoods approach: A framework for knowledge integration assessment. *Human Ecology Review, 13*(1), 90–99.

Korf, B. (2004). War, livelihoods and vulnerability in Sri Lanka. *Development and Change, 35*(2), 275–295.

Korf, B., & Oughton, E. (2006). Rethinking the European countryside: Can we learn from the South? *Journal of Rural Studies, 22*(3), 278–289.

Kotze, D. A. (2003). Role of women in the household economy, food production and food security: Policy guidelines. *Outlook on Agriculture, 32*(2), 111–121.

Krantz, L. (2001). *The Sustainable Livelihood approach to Poverty reduction: An Introduction.* Division for Policy and Socio-Economic Analysis: Swedish International Development Cooperation Agency.

Lapeyre, R. (2011). The Grootberg lodge partnership in Namibia: towards poverty alleviation and empowerment for long-term sustainability? *Current Issues in Tourism, 14*(3), 221–234.

Long, N. (1984). *Family and work in rural societies.* London: Tavistock Publications.

Lyons, M., & Snoxell, S. (2005). Creating urban social capital: Some evidence from informal traders in Nairobi. *Urban Studies, 42*(7), 1077–1097.

Mancini, F., Van Bruggen, A. H. C., & Jiggins, J. L. S. (2007). Evaluating cotton integrated pest management (IPM) farmer field school outcomes using the sustainable livelihoods approach in India. *Experimental Agriculture, 43*(1), 97–112.

Mathews, S. (2004). Post-development theory and the question of alternatives: A view from Africa. *Third World Quarterly, 25*(2), 373–384.

McLennan, B., & Garvin, T. (2012). Intra-regional variation in land use and livelihood change during a forest transition in Costa Rica's dry North West. *Land Use Policy, 29*(1), 119–130.

Meikle, S., Ramasut, T., & Walker, J. (2001). Sustainable urban livelihoods: concepts and implications for policy. Working Paper No. 112. London: University College

Mercer, J., & Kelman, I. (2010). Living alongside a volcano in Baliau, Papua New Guinea. *Disaster Prevention and Management, 19*(4), 412–422.

Mertz, O., Ravnborg, H. M., Lövei, G. L., Nielsen, I., & Konijnendijk, C. C. (2007). Ecosystem services and biodiversity in developing countries. *Biodiversity and Conservation, 16*(10), 2729–2737.

Millennium Ecosystem Assessment. (2005a). *Ecosystems and human well-being: Current state and trends.* Washington D.C.: Island Press.

Millennium Ecosystem Assessment. (2005b). *Ecosystems and human well-being: Synthesis.* Washington D.C.: Island Press.

Miller, D. B. (1977). Roles of naturalistic observation in comparative psychology. *American Psychologist, 32*(3), 211–219.

Morse, S. (2010). *Sustainability: A biological perspective.* Cambridge, UK: Cambridge University Press.

Morse, S., McNamara, N., Acholo, M., & Okwoli, B. (2000). *Visions of sustainability.* Stakeholders, change and indicators. Ashgate, Aldershot, UK.

Morse, S., & McNamara, N. (2006). Analysing institutional partnerships in development: A contract between equals or a loaded process? *Progress in Development Studies, 6*(4), 321–336.

Morse, S., & McNamara, N. (2008). Creating a greater partnership: Analyzing partnership in the Catholic Church development chain. *Area, 40*(1), 65–78.

Morse, S., & McNamara, N. (2009). The universal common good: Faith-based partnerships and sustainable development. *Sustainable Development, 17*(1), 30–48.

Moser, C. (1998). The asset vulnerability framework: Reassessing urban poverty reduction strategies. *World Development, 26*, 1–19.

Moser, G. G., Rogers, S., & van Til, R. H. (1997). *Nigeria: Experience with structural adjustment.* Washington DC: International Monetary Fund.

Mosley, P. (1992). Policy-making without facts: a note on the assessment of structural adjustment policies in Nigeria, 1985–1990. *African Affairs (London), 91*(363), 227–240.

Neefjes, K. (2000). *Environments and livelihoods: Strategies for sustainability.* Oxford: Oxfam.

Neylan, J. (2008). Social policy and the authority of evidence. *The Australian Journal of Public Administration, 67*(1), 12–19.

Nguthi, F. N., & Niehof, A. (2008). Effects of HIV/AIDS on the livelihood of banana-farming households in Central Kenya. *NJAS-Wageningen Journal of Life Sciences, 56*(3), 179–190.

Nha, T. (2009). Report of the FAO expert workshop on methods and indicators for evaluating the contribution of small-scale aquaculture to sustainable rural development. *FAO Fisheries and Aquaculture Technical Paper, 534*, 3–26.

Norberg, J. (1999). Linking Nature's services to ecosystems: Some general ecological concepts. *Ecological Economics, 29*(2), 183–202.

Odero, K. K. (2006). Information capital: 6th asset of sustainable livelihood framework. *Discovery and Innovation, 18*(2), 83–91.

Okwoli, P. E. (1973). *Short history of Igala*. Ilorin: Matanmi and Sons.

O'Neill, D. (2005). The promotion of ergonomics in industrially developing countries. *International Journal of Industrial Ergonomics, 35*(2), 163–168.

Pawson, R. (2006). Evidence-based policy: A realist perspective. Sage Publications

Perrings, C. (1994). Sustainable livelihoods and environmentally sound technology. *International Labour Review, 133*(3), 305–326.

Pieterse, J. N. (1998). My paradigm or yours? Alternative development, post-development, reflexive development. *Development and Change, 29*(2), 343–373.

Portes, A. (1998). Social capital: Its origins and applications in modern sociology. *Annual Review of Sociology, 22*, 1–14.

Rahnema, M., & Bawtree, V. (1997). *The post-development reader*. London: Zed Books.

Ravallion, M. (1997). Good and bad growth: The human development reports. *World Development, 25*(5), 631–638.

Rennie, K., & Singh, N. (1996). *Participatory research for Sustainable Livelihoods: A guidebook for field projects*. Manitoba, Canada: International Institute for Sustainable Development.

Roe, E. M. (1998). *Policy analysis and formulation for Sustainable Livelihoods*. New York: United Nations Development Programme.

Ruttan, V. W. (1984). Integrated rural development programmes: A historical perspective. *World Development, 12*(4), 393–401.

Sanderson, I. (2002). Evaluation, policy learning and evidence-based policy making. *Public Administration, 80*(1), 1–22.

Sanderson, D. (2009). Livelihoods approaches are a powerful tool for practice. London: DFID. Available at www.eldis.org/id21ext/SLA/SandersonJan09.html

Schuurman, F. J. (2000). Paradigms lost, paradigms regained? Development studies in the twenty-first century. *Third World Quarterly, 21*(1), 7–20.

Scoones, I. (1998). *'Sustainable Rural Livelihoods: A framework for analysis', Working Paper 72*. Brighton, UK: Institute for Development Studies.

Scoones, I., & Wolmer, W. (2003). Endpiece: The politics of livelihood opportunity. IDS Bulletin 34(3), 112–115

Seibel, H.D., & Damachi, U.G. (1982). Self help organisations. Guidelines and case studies for development planners and field workers: A participative approach. Bonn: Friedrich-Ebert-Stiflung.

Sen, A. K. (1984). *Resources, Values and Development*. Cambridge: Harvard University Press.

Sen, A. K. (1985). *Commodities and Capabilities*. Oxford: Oxford University Press.

Sen, A. K. (1999). *Development as freedom*. Oxford: Oxford University Press.

Serageldin, I., & Steer, A. (Eds.). (1994). *Making development sustainable: From concepts to action*. Washington DC: World Bank.

Sey, A. (2011). 'We use it different, different': Making sense of trends in mobile phone use in Ghana. *New Media and Society, 13*(3), 375–390.

Shankland, A. (2000). *'Analysing Policy for Sustainable Livelihoods', Research Report 49*. Brighton, UK: Institute of Development Studies.

Sidaway, J. D. (2007). Spaces of post-development. *Progress in Human Geography, 31*(3), 345–361.

Siddiqi, A. (2011). Supporting the working but vulnerable: Linkages between social protection and climate change. *Climate and Development, 3*(3), 209–227.

Siemiatycki, E. (2005). Post-development at a crossroads: Towards a 'real' development. *Undercurrent, 2*(3), 57–61.

Sillitoe, P. (2004). Interdisciplinary experiences: Working with indigenous knowledge in development. *Interdisciplinary Science Reviews, 29*(1), 6–23.

Simon, D. (2006). Separated by common ground? Bringing (post)development and (post)colonialism together. *The Geographical Journal, 172*(1), 10–21.

Simon, D. (2007). Beyond anti-development: Discourses, convergences, practices. *Singapore Journal of Tropical Geography, 28*(2), 205–218.

Simon, D., & Leck, H. (2010). Urbanizing the global environmental change and human security agendas. *Climate and Development, 2*(3), 263–275.

Singh, N., & Kalala, P. (1995). *Adaptive strategies and Sustainable Livelihoods: Community and policy studies for Burkino Faso, Ethiopia, Kenya, South Africa and Zimbabwe.* Manitoba, Canada: International Institute for Sustainable Development.

Singh, N., & Gilman, J. (1999). Making livelihoods more sustainable. *International Social Science Journal, 51*(162), 539–545.

Small, L. A. (2007). The sustainable rural livelihoods approach: A critical review. *Canadian Journal of Development Studies, 28*(1), 27–38.

Solesbury, W. (2003). Sustainable Livelihoods: A case study of the evolution of DFID policy. Working Paper 217. London: Overseas Development Institute.

Sorrell, S. (2007). Improving the evidence base for energy policy: The role of systematic reviews. *Energy Policy, 35*, 1858–1871.

Soyibo, A. (1996). *Financial linkage and development in Sub-Saharan Africa: The informal financial sector in Nigeria.* London: Overseas Development Institute.

Spiekermann, K., & Wegener, M. (2004) Modelling sustainable transport in the Dortmund Metropolitan Area. Paper presented at the STELLA FG4 meeting, Brussels, Belgium, 25–27 March 2004.

Srinivasan, T. N. (1994). Human development: A new paradigm or reinvention of the wheel? *The American Economic Review, 84*(2), 238–243.

Swift, J. (1989). Why are rural people vulnerable to famine? *IDS Bulletin, 20*(2), 8–15.

Tacconi, L. (2012). Redefining payments for environmental services. *Ecological Economics, 73*(1), 29–36.

Tao, T. C. H., & Wall, G. (2009). A livelihood approach to sustainability. *Asia Pacific Journal of Tourism Research, 14*(2), 137–152.

Tao, T. C. H., Wall, G., & Wismer, S. (2010). Culture and sustainable livelihoods. *Journal of Human Ecology, 29*(1), 1–21.

Tavakoli, M., Davies, H. T. O., & Thomson, R. (2000). Decision analysis in evidence-based decision making. *Journal of Evaluation in Clinical Practice, 6*(2), 111–120.

Tefera, T. L. (2009). Supply response, local reality and livelihood sustainability: The policy dilemma of khat (Catha edulis) production in eastern Ethiopia. *International Journal of Agricultural Sustainability, 7*(3), 176–188.

Toner, A., & Franks, T. (2006). Putting livelihoods thinking into practice: Implications for development management. *Public Administration and Development, 26*(1), 81–92.

Townsend, P., Phillimore, P., & Beattie, A. (1988). *Health and deprivation: Inequality and the North.* London: Croom Helm.

Turnhout, E., Hisschemoller, M., & Eijsackers, H. (2007). Ecological indicators: Between the two fires of science and policy. *Ecological Indicators, 7*, 215–228.

United Nations Development Programme (1990). Human Development Report 1990. New York: UNDP, Human Development Report Office.

Universities UK (2007). Talent wars: The international market for academic staff. Policy Briefing, July 2007, Universities UK, London. Available at www.universitiesuk.ac.uk

Uy, N., Takeuchi, Y., & Shaw, R. (2011). Local adaptation for livelihood resilience in Albay, Philippines. *Environmental Hazards: Human and Policy Dimensions, 10*(2), 139–153.

van Dillen, S. (2002). Book review: Rural livelihoods and diversity in developing countries. *Journal of Development Economics, 70*(1), 248–252.

Vatican II. The Documents 1963–1965. American Press and Associated Press. Also available at www.vatican.va/archive/hist_councils/ii_vatican_council/

Wlokas, H. L. (2011). What contribution does the installation of solar water heaters make towards the alleviation of energy poverty in South Africa? *Journal of Energy in Southern Africa, 22*(2), 27–39.

World Commission on Environment and Development (WCED). (1987). *Our Common Future.* Oxford: Oxford University Press.

Yudelman, M. (1976). The role of agriculture in integrated rural development projects: The experience of the World Bank. *Sociologia Ruralis, 16*(3), 308–325.

Zoomers, A. (2005). Three decades of rural development projects in Asia, Latin America, and Africa: Learning from successes and failures. *International Development Planning Review, 27*(3), 271–296.

Index

S. Morse and N. McNamara, *Sustainable Livelihood Approach*,
DOI: 10.1007/978-94-007-6268-8,
© Springer Science+Business Media Dordrecht 2013

Printed by Printforce, the Netherlands